Seabird Ecology

TERTIARY LEVEL BIOLOGY

A series covering selected areas of biology at advanced under-graduate level. While designed specifically for course options at this level within Universities and Polytechnics, the series will be of great value to specialists and research workers in other fields who require a knowledge of the essentials of a subject.

Recent titles in the series:

Locomotion of Animals	Alexander
Animal Energetics	Brafield and Llewellyn
Biology of Reptiles	Spellerberg
Biology of Fishes	Bone and Marshall
Mammal Ecology	Delany
Virology of Flowering Plants	Stevens
Evolutionary Principles	Calow
Saltmarsh Ecology	Long and Mason
Tropical Rain Forest Ecology	Mabberley
Avian Ecology	Perrins and Birkhead
The Lichen-Forming Fungi	Hawkesworth and Hill
Plant Molecular Biology	Grierson and Covey
Social Behaviour in Mammals	Poole
Physiological Strategies in Avian Biology	Phillips, Butler and Sharp
An Introduction to Coastal Ecology	Boaden and Seed
Microbial Energetics	Dawes
Molecule, Nerve and Embryo	Ribchester
Nitrogen Fixation in Plants	Dixon and Wheeler
Genetics of Microbes (2nd edn.)	Bainbridge
The Biochemistry of Energy Utilization in Plants	Dennis

TERTIARY LEVEL BIOLOGY

Seabird Ecology

R.W. FURNESS, BSc, PhD
Lecturer in Zoology
University of Glasgow

P. MONAGHAN, BSc, PhD
Lecturer in Zoology
University of Glasgow

Blackie

Glasgow and London

Published in the USA by
Chapman and Hall
New York

Blackie & Son Limited
Bishopbriggs, Glasgow G64 2NZ
7 Leicester Place, London WC2H 7BP

Published in the USA by
Chapman and Hall
in association with Methuen, Inc.
29 West 35th Street, New York, NY 10001

British Library Cataloguing in Publication Data

Furness, R. W.
 Seabird ecology.—(Tertiary level biology)
 1. Sea birds—Ecology
 I. Title II. Monaghan, P. III. Series
 598.252'636 QL673

 ISBN 0-216-92087-6
 ISBN 0-216-92088-4 Pbk

Library of Congress in Publication Data

Furness, R. W.
 Seabird ecology.

 (Tertiary level biology)
 Bibliography: p.
 Includes index.
 1. Sea birds—Ecology. 2. Birds—Ecology.
 I. Monaghan, P. II. Title III. Series
 QL673.F87 1987 598.29'162 86-20698
 ISBN 0-412-01451-3
 ISBN 0-412-01461-0 (pbk.)

Photosetting by Digital Publications Ltd., Edinburgh, Scotland.
Printed in Great Britain by Bell & Bain (Glasgow) Ltd.

Preface

In the last few years there has been an exciting upsurge in seabird research. There are several reasons for this. Man's increased exploitation of natural resources has led to a greater awareness of the potential conflicts with seabirds, and of the use of seabirds to indicate the damage we might be doing to our environment. Many seabird populations have increased dramatically in numbers and so seem more likely to conflict with man, for example through competition for food or transmission of diseases. Oil exploration and production has resulted in major studies of seabird distributions and ecology in relation to oil pollution. The possibility that seabirds may provide information on fish stock biology is now being critically investigated. Some seabird species have suffered serious declines in numbers and require conservation action to be taken to reduce the chances that they will become extinct. This requires an understanding of the factors determining their population size and dynamics.

This book is the first to deal in depth with the relationships between seabirds and man, the problems that these present and the ways in which we can attempt to solve these problems to our benefit and to the long-term benefit of seabirds and the marine environment. Many of the aspects described in this book have yet to be fully understood, never mind solved. We have aimed to provide a text that will be a useful case study in applied biology, as well as giving an insight into some important issues in marine biology, and we have pointed to some exciting areas for future research in the ecology of seabirds, interactions with fisheries, environmental monitoring, pest control and conservation. This book should therefore be of considerable interest to undergraduate and postgraduate students in the fields of applied biology, marine biology and ecology, and to serious amateur ornithologists who wish to know more about the ecology of the seabirds that they watch.

We would particularly like to thank Dr Neil Metcalfe and Dr Susan Anderson for considerable assistance with drawing of figures and helpful suggestions for improving the text. Dr Sandra Muirhead and Miss Anne Hudson provided us with helpful comments on particular topics and allowed us to quote from unpublished data. Bernard Zonfrillo and Mrs Elizabeth Denton also assisted with preparation of figures, and parts of the manuscript were typed by Mrs M. McCulloch and Mr M.L.N. Murthy. Kenneth Ensor helped with proof reading and the preparation of the index.

RWF

PM

Contents

Chapter 5. INTERACTIONS WITH FISHERIES   53

Chapter 6. MONITORING MARINE ENVIRONMENTS   100

CHAPTER ONE

INTRODUCTION

The sea is a hostile, variable environment. A comparatively small number of birds is adapted to life in this habitat despite the enormous amount of food potentially available in surface waters; only 3% of the 8600 known species of bird exploit this two-thirds of the world's surface. Man has taken up the challenge of obtaining resources from the sea, and in so doing interacts with seabirds as a partner, predator and competitor.

The first seabird observers were mariners, and a working knowledge of seabird identification and the distances various species travel from land was an important part of the early navigator's art (Brown, 1980). Prior to the availability of sonar methods, fishermen used seabirds in the location of fish schools, and Japanese Cormorants are still employed to catch fish in traditional Japanese fisheries. Outside the breeding season, seabirds are generally too dispersed to represent an economically exploitable resource to man. However, where conditions are favourable, vast breeding concentrations of particular species can occur, often involving millions of birds. These have been utilized in a variety of ways by human populations. Where large breeding colonies occur in dry climates, the droppings of seabirds collect to such an extent that they can be mined for use as fertilizer, as in Peru where this 'guano' is an integral part of the local economy. The availability of seabird eggs, chicks and adults collected for food and oil has been important to many communities, and until recently, permanent habitation of many otherwise barren offshore islands would not have been possible without this resource. Various specialized harnesses and pulleys were devised to reach inaccessible nest sites, often at considerable risk to the user. However, seabirds were not only killed to obtain much needed

1

food. During the nineteenth century, adults and young were shot in huge numbers both for sport and to supply feathers for use in the millinery trade. An outcry in Britain and the United States over this senseless slaughter gave rise to the world's first conservation movement. The present century has seen a considerable change in attitude to seabirds, from one of exploitation to conservation, and many seabirds have increased in numbers. It is somewhat ironic that this population recovery is now accompanied by concern that seabirds may represent competitors to our fishing industries.

This book deals with these interactions between man and seabirds. In so doing, it illustrates that, if we are to co-exist with other animals, and successfully manage and conserve resources and environments, we must understand the behaviour and ecology of the species concerned.

The next chapter examines seabird lifestyles, many aspects of which have influenced their association with human populations. The type of breeding site used and the fact that seabirds typically nest in dense colonies determined the extent to which they could be economically exploited. Clearly, an understanding of their life history strategies and population regulation (Chapter 4) is essential if we are to predict how they will respond to environmental changes that we bring upon them. Chapter 3 outlines the ways in which seabirds forage and the types of prey they take, necessary information in evaluating the extent to which they compete with us for marine resources, and in our attempts to conserve endangered species. In addition, the water depth at which seabirds forage influences the extent to which they become caught in fishing gear, their diving behaviour influences their susceptibility to oiling, and their choice of feeding site influences their role in the spread of disease. Chapters 5, 6, and 7 deal in depth with the interactions between seabirds and man with respect to fisheries, pollution of marine environments and pest problems caused by those seabirds which have benefited from their capacity to exploit our waste products. The final chapter examines seabird conservation requirements.

By understanding their relationship with the marine environment, and the impact of our interactions with these birds, our continued association with seabirds need not give rise to the kind of fate which recently befell the Great Auk, driven to extinction in 1844 by excessive and relentless human persecution.

CHAPTER TWO

SEABIRD LIFE STYLES

2.1 What is a seabird?

'Seabird' is a rather loose term traditionally used to cover those birds which obtain at least part of their food from the sea, not simply by wading into it as do shorebirds, but by travelling some distance over its surface; in addition, they typically breed on offshore islands or coastal areas. The main seabird groups fall into four taxonomic orders, involving some 274 species, summarized in Table 2.1. Certain other birds, such as seaducks, divers, grebes and phalaropes are also sometimes classed as seabirds.

Some seabird species generally feed in water within sight of land, and are termed 'inshore feeders', while others, the 'offshore feeders', feed out of sight of land. (A component of the latter do

Table 2.1 The main taxonomic groups of seabirds (after Harrison, 1983)

ORDER	FAMILIES	No. of species	Common names
SPHENISCIFORMES	Spheniscidae	16	Penguins
PROCELLARIFORMES	Diomedeidae	13	Albatrosses
	Procellariidae	55	Fulmars, prions, petrels, shearwaters
	Hydrobatidae	20	Storm petrels
	Pelecanoididae	4	Diving petrels
PELECANIFORMES	Phaethontidae	3	Tropicbirds
	Pelecanidae	7	Pelicans
	Phalacrocoracidae	27	Cormorants, shags
	Fregatidae	5	Frigate-birds
CHARADRIIFORMES	Stercorariidae	6	Skuas
	Laridae	87	Gulls, terns, noddies
	Rynchopidae	3	Skimmers
	Alcidae	22	Auks

3

Figure 2.1 Typical breeding sites used by some North Atlantic seabirds.

not return to land at night outside of the breeding season, and are often termed 'pelagic'.) Seabirds tend to be larger than landbirds and are generally much less colourful. Their body plumage is some combination of black, white, brown and grey, and there is little sexual dimorphism. Such colourful features as do occur are confined to the head, facial skin, bill and feet. There has been considerable discussion of the function of the monochromatic coloration of seabirds and, while there is some evidence that birds with white underparts are not reacted to as quickly by fish, social signalling may also be involved.

2.2 Habitat selection

Habitat selection is essentially the choice of a place in which to live (Partridge, 1978). All seabirds are dependent on land to breed, and food resources, microclimate, topography, substrate, availability of nest material, prevalence of ectoparasites and the presence both of conspecifics and other species are all known to influence the choice of a particular site (Buckley and Buckley, 1980). Seabirds generally breed in comparatively inaccessible coastal areas, and the typical sites used by some North Atlantic seabird species are shown in Figure 2.1. The use of cliffs for breeding protects vulnerable eggs and young from ground predators; where such predators are absent, as for example on South Georgia Island in the sub-antarctic, cliff nesting is much less common (Croxall and Prince, 1980).

Variation in habitat selection also occurs within species, and this tends to be especially marked in those species which have shown rapid population increases and expansion into new nesting habitats. Herring Gulls, for example, which have recently expanded rapidly throughout much of their range, use a wide variety of breeding sites, including coastal cliffs and rocky outcrops, grassy slopes, boulder and rocky beaches, sand dunes, shingle, moorland, trees (rarely) and even inhabited buildings (Figure 2.2).

The quality of the nest site can have considerable effects on breeding performance, as has clearly been demonstrated in shags breeding on a coastal island off the north-east coast of England. An exceptionally high mortality of adults in 1968, caused by ingestion of a neurotoxin produced by a bloom of the protozoan *Gonyaulax* (a so-called 'red tide'), had the result that nest sites were no longer in short supply in the following years. As a consequence, the average

Figure 2.2 Herring Gulls breeding on an inhabited building. The number of Herring Gulls breeding in towns in Britain has increased considerably during this century (see 7.3.1).

nest site quality of breeding shags increased in 1969, as did their breeding success; this was especially marked in first time breeders which moved into high quality sites from which they would normally have been excluded by established breeding adults (Potts *et al.*, 1980).

2.3 Mating systems

Seabirds are almost invariably monogamous. This is presumably because successful incubation and chick rearing by most marine birds demands the continuous attention and co-operation of both parents. Thus, males cannot increase their reproductive output by taking more than one mate, and in seabirds polygyny is not favoured by natural selection. Individual male skuas have occasionally been reported breeding simultaneously with two females, but both females generally have a reduced breeding success (Hunt, 1980). Therefore, to maximize reproductive success, males and females must try to attract a mate of as high a quality as they possibly can.

At the large crowded colonies where seabirds typically nest (see section 2.3), birds may either establish a nesting site and attempt to attract a mate, or gather together at so-called 'club' sites, separate from, but close to, the breeding areas. The latter commonly occur in gulls, but are otherwise comparatively rare; most seabirds pair at the nest site, though some (for example Sandwich Terns) tend to arrive at the breeding colonies after pair formation has taken place (Veen, 1977). Assessment of mate quality can involve complex courtship displays, such as the knocking together of bills in Puffins (Harris, 1984) or the astonishing dances of albatrosses, engagingly described by Nelson (1980). Females will also examine the potential nest sites and/or territories secured by prospective mates.

In some seabird groups, in particular the gulls, terns and skuas, feeding of the female by the male prior to egg laying may be a significant component of mate choice. During such courtship feeding, the female can obtain information on the size and quality of food the male can obtain, which may be correlated with his capacity to feed chicks later in the breeding cycle. In common terns, males which are seeking a mate carry fish around the colony and display to unmated females, but do not feed them regularly until the pair becomes established. It is during this early phase of courtship that many incipient pairs break up (Nisbet, 1973). Courtship feeding may also provide the female with energy for egg production, which may be very important in gulls and terns, since the females of these species produce a comparatively expensive clutch, the three eggs together comprising some 29% of their body weight, considerably more than in most other groups (Hunt, 1980). Nisbet found that male Common Terns provided females with a substantial amount of food, particularly just before egg laying when the female may have difficulty in foraging efficiently due to the added burden of developing eggs. The courtship feeding performance of male Common Terns correlated well with the total weight of the clutch their mates produced, and to a lesser extent with subsequent fledging success (Nisbet, 1973).

Most seabirds tend to pair with individuals of a similar age to themselves, and provided the pair are successful, they will continue to breed together while both remain alive (Hunt, 1980). However, they will change mate and/or nest position if breeding is not successful. In the Kittiwake, retention of the same mate has been shown to result in earlier breeding and improved fledging success, and it has been similarly shown for Atlantic Gannets, Fulmars, Manx Shearwaters and Arctic Skuas that the longer a pair have bred together, the better their breeding success. In birds breeding for the first time, pairing with an experienced mate may improve breeding performance (Coulson and Thomas, 1985a).

2.4 Colonial breeding

Colonial breeding is common in birds, and recent general reviews of its adaptive significance can be found in Perrins and Birkhead (1983) and Wittenberger and Hunt (1985). Some 98% of the 274 seabird species typically nest in colonies, and colony size can range from only a few breeding pairs to over a million birds. While a number of seabirds occasionally nest solitarily, especially the large gulls when feeding inland, only two species of murrelets and one gull do so habitually. Clearly, colonial breeding must be especially advantageous for marine birds.

Nesting in colonies is likely to have both costs and benefits, and these may be traded off one against the other. Putative costs and benefits of breeding at central localities can be summarized under three broad headings, those relating to foraging, predation and social behaviour. The selection pressures which have favoured colonial breeding are undoubtedly not the same for different species. For some, the main driving force may have been a shortage of safe breeding sites, while for others, effective predator defence may have been more important. Once colonial breeding becomes established, however, effects such as limitation of breeding density may come into play. Such consequences of colonial breeding must not be misconstrued as functions.

We shall now consider the potential costs and benefits of colonial breeding in seabirds.

2.4.1 *Foraging*

Foraging benefits from colonial breeding in seabirds could result if birds are able to obtain information on the whereabouts of food supplies by following successful conspecifics to good feeding places (Ward and Zahavi, 1973). There is little evidence to suggest that this occurs at seabird colonies (Andersson *et al.*, 1981; Wittenberger and Hunt, 1985). Foraging disadvantages may relate to increased competition for food around the colony, and there is some evidence that this may occur. In the Pribilof Islands off the Alaskan coast for example, Thick-billed Murres have lower chick growth rates and lower fledging weights on an island where the population numbers 1.5 million, compared with those on another island where the breeding population is only 150 000 (Wittenberger and Hunt, 1985). The relationship between seabird food supplies and population size is discussed in more detail in Chapter 4.

2.4.2 *Predation*

Seabirds breeding in colonies may be better able to defend their eggs and chicks against predators by effective communal mobbing. This is certainly the case in gulls, whose dive-bombing behaviour at large colonies can successfully deter many ground predators. With respect to predators which take only one prey per foraging trip, the risk of individual birds or their offspring being the victim of an attack is reduced if large numbers of potential prey are present (Hamilton, 1971); the risk to young birds is further reduced if breeding is synchronized. Alternatively, colonial breeding may arise because it is advantageous for seabirds to breed at sites which are safe from predators and, since such sites may be in short supply, birds collect together where they occur.

A disadvantage of breeding in colonies is that the colony may be very conspicuous and therefore easy for predators to locate. Moreover, if the predator takes more than one prey at a time, it has found an abundant food source on which to concentrate. It is for just such reasons that seabird colonies can constitute an economically exploitable resource to humans.

2.4.3 *Social behaviour*

The social advantages of colonial breeding in seabirds are more complex and somewhat more speculative. A colony may serve as a meeting place for potential mates, and a high, local density of displaying conspecifics may serve to accelerate and synchronize the onset of egg laying (the so-called 'Fraser-Darling' effect). While Kittiwakes, for example, do not appear to be able to breed without at least some social stimulation from neighbouring pairs (Coulson and Dixon, 1979), most of the effects of colony density on timing of breeding reported to occur in gulls appear to result from the fact that low-density colonies contain mostly young birds, which tend to

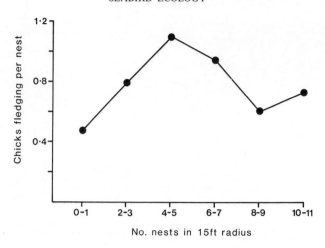

Figure 2.3 The relationship between nesting density and breeding success in Herring Gulls. Birds breeding at intermediate densities had the highest breeding success. (Data from Parsons, 1976.)

lay comparatively late in the season. In other words, the effect of nesting density on laying date was confused with the effect of age.

Nesting beside conspecifics can give rise to many deleterious effects on breeding success. This can come about through increased competition, disease transmission, misdirected parental care due to brood mix-ups, and the killing of young either in territorial disputes or by cannibalistic adults. Parsons (1976) found a clear relationship between breeding density and reproductive success in Herring Gulls; independent of laying date, those nesting at intermediate densities had the highest average clutch sizes, hatching success and fledging success, and consequently the highest overall breeding success (Figure 2.3). This was because birds nesting at low densities were selectively predated by cannibals, while those at high densities lost many young in territorial disputes. When Herring Gulls nest in towns, they can in fact have a higher than normal breeding success due to the effective low nesting density enforced on the birds by the typical layout of buildings; they can nest close to each other, but access between territories is restricted by roof structure and fewer young are killed by neighbouring birds (Monaghan, 1979).

Competition from conspecifics for mates or breeding sites may also occur. In African Cormorant and tropicbird colonies a shortage of nest sites prevents all individuals from breeding at the same time. Pairs must delay breeding until a suitable site is vacated (Marshall and Roberts, 1959; Snow, 1965). Even where breeding sites do not apparently differ in predation risk, social effects on breeding success can still operate. Kittiwakes nesting in the centre of a uniform colony which was free from predation reared 11% more chicks than those nesting on the edges (Coulson and Thomas, 1985a). Central areas of seabird colonies appear to be particularly attractive to young birds prospecting for nesting sites; these birds try, albeit unsuccessfully, to obtain sites in such areas rather than on the colony periphery, and may delay breeding as a result of the high competition from established adults (Chabrzyk and Coulson, 1976). Colony distribution and density will therefore be constrained by such social factors. A further disadvantage which may result from colonial breeding arises from the increased opportunities for extra-pair copulations (Birkhead *et al.* 1985), and the consequent increased risk to males of rearing offspring to which they are not related.

2.5 Life-history theory

During their lives, animals need to divide their time, and the resources they obtain from their environments, between staying alive and producing offspring. However, a high input into breeding may increase the risk of dying. The ways in which animals allocate resources to these potentially conflicting demands of survival and reproduction are referred to as their life-history strategies. Some species have a comparatively high reproductive rate and short lifespan, while others reproduce at a much lower rate, but do so over several breeding seasons (though there is no simple dichotomy between the two tactics). Life history theory attempts to explain such different strategies in the light of ecological conditions; the mortality patterns of adults and juveniles, and the degree of intraspecific competition, seem to be particularly important (Horn and Rubenstein, 1984). High juvenile mortality, coupled with high levels of competition, tends to favour animals' conserving resources for themselves rather than their offspring and thereby surviving longer; investing in only a few well provided young whose chances of survival and competitive abilities are enhanced is also favoured. Species showing this pattern are often referred to as being *K*-selected, reflecting the fact that they tend to live in stable populations whose size is at or near the environmental carrying capacity (represented by the parameter K in equations of population growth). Selection will therefore favour the ability to compete with conspecifics in comparatively crowded environments. On the other hand, high adult mortality, and an environment in which animals are not much constrained by intra-specific competition, seem to favour the production of many small young; this is done even at the expense of parental survival, since the chances of individuals surviving to reproduce again are low in any case. Such species are termed r-selected, since they tend to have high per capita rates of population growth (represented by the parameter r in equations of population growth), but their population size is subject to considerable fluctuations.

In general, seabirds represent extreme K-selected species. Adult survival is generally high, and annual reproductive output low. Figure 2.4 shows the main features of this life-history strategy, and also the main positive feedback loops whereby it is reinforced.

2.6 Comparative seabird life histories

Within this general K-selected life-history pattern, there is some variability in the way in which different seabird species allocate resources to survival and reproduction. Several authors (e.g. Lack,1968; Ashmole, 1971; Nelson, 1983) have tried to relate this variability to ecological conditions, in particular food availability. Species which feed well offshore tend to have higher foraging costs than those which feed near their breeding sites, and during the breeding season marine food tends to be more abundant in high latitudes than it is in tropical waters. However, it is difficult to separate the effects of foraging distance and breeding latitude, since most tropical seabirds tend to feed offshore, and these variables have been further confused with other factors such as differences in body size.

Body size is certainly an important consideration in annual energy budgets. Energetic costs per unit size are higher in small than in large birds; the costs of staying alive are therefore proportionately less in larger species. Time taken to produce young will also be related to the size to which they must grow. In addition to these size effects, energy expended in survival and reproduction will vary in relation to flying and foraging techniques, and to anatomical structure, such as the size and shape of the wing relative to body size, which affects the energetic costs of flying.

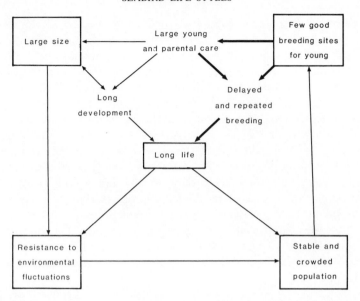

Figure 2.4 The main features of seabird life history strategies, and the positive feedback loops where by this pattern tends to be reinforced. The heavy arrows represent natural selection. The survival of young is improved by parental care and long development times. This slow maturation rate is also influenced by the high adult survival rates, which means that young must compete with adults for breeding places. All of these factors favour large body size, which makes the animals fairly resistant to environmental changes, but also needs a long development time. (Based on Horn and Rubenstein, 1984.)

Different seabird groups may not solve similar environmental problems in the same way, and we may therefore find life-history differences between species which relate as much to the design of the birds (a consequence of phylogenetic history) as to prevailing ecological conditions. For example, all procellariids (Fulmars, prions, petrels and shearwaters) lay one egg, despite the breeding range of this group extending from the tropics to high latitudes, and considerable differences in feeding ecology (Cramp 1977).

We shall now examine some seabird life history components in more detail.

2.6.1 *Longevity*

It is very difficult to obtain data on survival rates for long-lived species in the wild. There is no known way of ageing birds once they have obtained adult plumage, and most estimates of survival rely on the reporting of ringed birds found dead by the general public. Such methods are fraught with difficulties. For example, some birds lose their rings, and finding and reporting rates vary between species, geographical localities and at different times of year. What we know about seabird mortality rates in general tells us that they tend to be very long-lived birds indeed. Table 2.2 gives estimated average annual mortality rates for a number of species, plus the adult life expectancy and typical age of first breeding. It has been suggested that inshore feeders tend to have longer lives than offshore feeders (Lack, 1968), but more recent data do not support this (Table 2.2).

An important aspect of most methods of calculating survival rates based on birds ringed as young is the assumption that the survival rate of birds, unlike most other animals, does not vary with age once maturity has been reached (Lakhani and Newton, 1983); in other words, that birds do not become senile! While this widely held belief may be true of birds which have a very short lifespan, such as most small songbirds, it is probably not the case for some long-lived seabirds. Using observations on individually colour-marked Kittiwakes, Coulson and Wooller (1976) found that these birds showed a progressive decrease in survival rate with increasing age. This was especially marked in females, those which had bred more than seven times having twice the mortality rate of females which had bred only once. Coulson and Wooller also found that survival in the Kittiwake varies with sex and nesting position in the breeding

Table 2.2 The average age of first breeding, annual adult mortality rate and length of breeding life in some inshore- and offshore-feeding seabirds in the North Atlantic. Data from Cramp (1978 and 1985) and Cramp and Simmons (1983).

Feeding Range	Species	Average age of first breeding (years)	Average annual adult mortality	Average breeding life (years)
Onshore	Common Tern	3–4	8.0%	12.0
	Arctic Tern	4	13.0%	7.2
	Razorbill	5	10.0%	9.5
	Puffin	5–6	4.0%	24.5
	Herring Gull	5	6.0%	15.0
Offshore	Fulmar	6–12	5.5%	17.7
	Manx Shearwater	4–5	10.0%	9.5
	Storm Petrel	4–5	12.0%	7.8
	Gannet	5–6	6.0%	16.2

colony; males had higher mortality rates than females, and males breeding in the centre of the study colony tended to live longer than those breeding on the edge, particularly during the early years of colony formation.

2.6.2 *Age of first breeding*

There is some tendency for the young of offshore feeding seabirds to start breeding later in life than those of inshore feeders, and this appears to be true within and between seabird groups (Table 2.2). For example, the longest period of immaturity in gulls and terns is found in the offshore-feeding Sooty Tern, and virtually all pelecaniforms which forage far from their breeding colonies delay breeding until five or more years old (Nelson, 1983). The period of immaturity before breeding begins will in part reflect the time taken to aquire complex foraging skills necessary for successful breeding (see Chapter 3), in addition to possible social constraints (see 2.3.3), and young inexperienced birds often have comparatively poor breeding success (Hunt, 1980; Ryder, 1980).

Since reproductive effort may be costly (see 2.5.4) breeding at an early age will not necessarily increase lifetime reproductive success; in Fulmars, for example, birds breeding for the first time at six years old have higher subsequent mortality rates, and thereby shorter breeding lives, than those which delay breeding until they are nine years of age or older (Ollason and Dunnet, 1978).

2.6.3 *Timing of breeding*

The breeding season itself is usually timed to coincide with periods of maximum food availability, and costly maintenance activities such as feather moult may be incompatible with the demands of reproduction. At high latitudes, the period of maximum food availability tends to be short in comparison with the less seasonal environment in the tropics where seabird breeding is correspondingly less synchronized and extends over a longer period. Common Cormorants, for example, breed only in late spring and summer in Europe, but breeding occurs more or less continuously in tropical Australia (though individual birds breed only once per year). In some tropical seabirds such as the Sooty Tern and the Swallow-tailed Gull, breeding and moulting can be completed in less than 12 months, and individual birds breed at less than annual intervals (Ashmole, 1971). In contrast, some large seabirds such as Abbott's Booby and the Magnificent Frigate-bird in the tropics, and the Wandering Albatross and King Penguin at high southern latitudes, take more than a year for each successful breeding attempt; they can therefore breed successfully only every other year. These latter Antarctic species therefore have to breed during the intensely cold Antarctic winter. The most astonishing feat of endurance is performed by the Emperor Penguin, which breeds on the Antarctic sea ice. These remarkable birds lay their eggs in May, during the dark Antarctic winter. After laying, the female departs to feed while the male incubates the egg with his feet until she returns (Jouventin, 1975); he must wait over two months without food at temperatures below $-35°C$!

2.6.4 *Clutch and brood size*

The size of clutch produced by birds is generally held to be that which maximizes lifetime, though not necessarily annual, reproductive output. Nearly all offshore feeding seabirds lay single-egg clutches, whereas most inshore feeders have clutches of two or three. The comparatively high energetic and time costs of foraging offshore may limit clutch size (Lack, 1968), and many offshore feeders such as shearwaters appear to be unable to rear more than one young in a single season. In contrast, it has been demonstrated that some seabirds, for example North Atlantic Gannets and several gull species, are capable of rearing more young in a season than they typically do (Nelson, 1978; Winkler, 1983). However, there is likely to be a trade-off between reproductive effort and survival to breed again another year; increased effort in one season may carry an increased mortality risk. There are few relevant data on this subject. It has been shown in King Penguins that the effort of breeding increases mortality rates (Stonehouse, 1960). But this is not always the case. Those female Kittiwakes which lay larger than normal clutches (three rather than two eggs) rear more young per season and also appear to live longer! This may indicate that these are comparatively high quality individuals (Coulson and Porter, 1985).

Where two or more eggs are laid and the chicks hatch, sibling aggression may reduce brood size when food is in short supply, as occurs in South Polar Skuas (Evans, 1980). Masked and Brown Boobies lay either one or two eggs, but two-chick broods are always

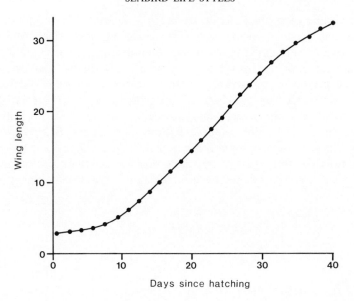

Figure 2.5 Typical S-shaped growth curve of a Herring Gull chick.

reduced to one by such 'siblicide', regardless of food abundance (Nelson, 1978). It will generally be the fitter chick which survives. Eudyptes penguins are unique amongst birds in laying two eggs of markedly different sizes; the first is very small and often does not hatch. In one of these species, the Rockhopper Penguin, the small egg often does hatch. If the larger egg fails to hatch this small chick is sometimes reared, but otherwise it is ignored (Warham, 1963).

2.6.5 *Chick growth and parental care*

Reproductive effort is also related to the rate at which the young must be fed. Food requirements of chicks will vary with growth rates, the amount of fat deposition, the cost of thermoregulation, the degree of activity and other maintenance costs (Dunn, 1979). Chick growth rates are generally estimated by periodically weighing chicks of known age. The resulting curve tends to be S-shaped (logistic), as shown in Figure 2.5; growth rates can either be expressed as those during the linear phase of the growth period, or by converting the curve to a straight line using appropriate conversion methods as outlined by Ricklefs (1967).

The young of offshore feeders tend to grow more slowly than those of inshore feeders, even when differences in body size are taken into account. Figure 2.6 shows the relationship between adult body weight and fledging period for a number of Atlantic seabirds. The fledging period is considerably longer for offshore feeders compared with inshore feeders of a similar size, breeding at similar latitudes (e.g. the Manx shearwater and the Puffin, Leach's Petrel and the Little Tern). There are some exceptions such as the Fulmar, whose single chick grows comparatively quickly (Figure 2.6).

To examine the effect of breeding latitude on chick growth rates, it is necessary to examine birds of similar sizes and feeding ranges

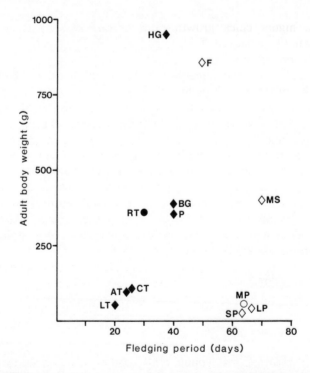

Figure 2.6 The relationship between adult body weight and fledging period for a number of North Atlantic onshore ◆ and ◊ offshore feeding seabirds. Also shown for comparison are two tropical breeding seabirds, the onshore feeding Royal Tern and the offshore feeding Madeiran Storm Petrel. HG = Herring Gull; BG = Black Guillemot; P = Puffin; CT = Common Tern; AT = Arctic Tern; LT = Little Tern; RT = Royal Tern; F = Fulmar; MS = Manx Shearwater; LP = Leach's Storm Petrel; SP = British Storm Petrel; MP = Madeiran Storm Petrel.

breeding in tropical and temperate areas. There appears in fact to be little latitudinal effect on growth rates—compare the inshore feeding Royal Tern (tropical breeder) and Puffin (temperate breeder), and the offshore feeding Madeiran Storm Petrel (tropical breeder) and British Storm Petrel (temperate breeder) (Figure 2.6). Foraging range seems to have more effect on growth rates than breeding latitude.

Offshore feeders tend to feed their chicks at intervals of 2–3 days, as compared with the typical pattern of several feeds per day of inshore feeders (Lack, 1968). To some extent, provisioning will be related to the demands of the chick (Ricklefs *et al.*, 1985) and predation risks to adults, but the overall pattern of chick development is likely to be such that reproductive costs for the parent are not unduly high. Demonstrating that offshore species have the capacity to sustain higher chick growth rates in one season, as shown in the tropical Grey-backed Tern (Shea and Ricklefs, 1985) does not necessarily prove that the development rate of the chick has not been shaped by natural selection to maximize parental survival and thereby the number of breeding attempts.

The young of many offshore feeders have a considerable fat store which may aid survival during periods of food shortage. In the offshore feeding King Penguin, the young have a large amount of body fat to draw on during the winter period, during which they are fed by their parents only once every five or six weeks. Most seabird species continue to feed the chick until it nears fledging, and some, such as Razorbills and Guillemots, reduce the cost of foraging by taking the young chick to the food source, which often means it must leap off a cliff several hundred feet high to reach the sea, before it can actually fly!

After fledging, young seabirds usually forage independently of their parents, though some gulls and boobies continue feeding their young for several weeks after fledging (Burger, 1980).

2.6.6 *Dispersal and migration*

In most seabird species, young birds tend to disperse further from the breeding colonies during their non breeding periods than do adults, though many young eventually return to their colony of birth to breed (philopatry). Once having bred in a particular locality, most adults are very site tenacious, that is, they return to that area in subsequent breeding seasons. Outside the breeding season, seabirds are no longer tied to land, and most disperse widely over seas and oceans to areas of high food availability. For those breeding at high latitudes, this may involve a migration to the opposite hemisphere to escape harsh winter conditions. The Arctic Tern for example migrates from its Arctic and boreal breeding grounds to the edge of the Antarctic pack ice. Such migrations can be very rapid indeed, as illustrated by a juvenile Manx Shearwater, recovered on the coast of Brazil 17 days after being ringed at its birthplace in Wales; it had travelled 9500 km, at an average speed of 650 km per day (Spencer, 1966). Other species such as gulls and cormorants may undertake only short migrations, and in some areas are more or less resident.

Long-distance migration by seabirds, as in all birds, is a risky undertaking, and is only likely to occur where environmental conditions prevent continuous occupation of the same area throughout the year.

2.7 Conclusions

In comparison with other types of bird, seabirds are generally long-lived, lay small clutches and delay breeding until at least the second year of life, indeed usually much longer. They are generally monogamous, colonial breeders, with a strong tendency to return to their colony of birth to breed. Variations in seabird life history patterns are related to some extent to different foraging patterns, and thereby to the costs of breeding.

It is interesting that many of the factors which are held to favour the evolution of co-operative breeding in birds (where young birds assist others, usually their parents, in breeding rather than breeding themselves), such as philopatry, site tenacity and delayed maturity, are present in seabirds, yet co-operative breeding is rare if it occurs at all (Perrins and Birkhead, 1983; Emlen, 1984). However, co-operative breeding is in addition favoured by a severe shortage of breeding territories or mates; in helping their parents, young birds may increase their chances of inheriting the breeding territory or obtaining a mate, while at the same time increasing their inclusive fitness through kin selection if they are helping their parents. Breeding sites *per se* are rarely limiting in most seabird species. Young seabirds presumably do better in terms of lifetime reproductive success by breeding themselves as soon as they are competent, rather than helping their parents.

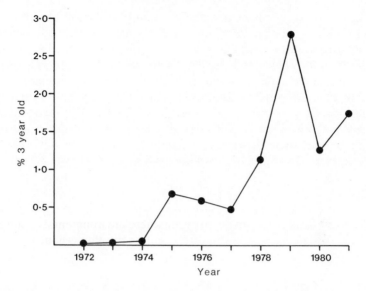

Figure 2.7 Changes in the number of three-year-old Herring Gulls breeding on the Isle of May, Scotland, from 1972 to 1981. During this period the colony decreased greatly in size, due to culling of breeding birds. (Based on Coulson *et al.*, 1982.)

2.8 Interactions with man

An understanding of seabird behaviour and life-history strategies is essential if we are to predict the ways in which their populations will respond to human exploitation or environmental catastrophies, and if we are to design appropriate management policies. For example, indiscriminate culling may, by reducing population density, produce rather surprising effects. Large-scale culling of Herring Gulls for management purposes in a colony in the Firth of Forth has maintained the breeding birds at an artificially low density. As a result, a higher proportion of the young gulls are now breeding at three years rather than five (Figure 2.7) since they can more easily obtain nesting sites in the colony; an increased proportion of young are also returning to this colony to breed rather than moving elsewhere. In addition, the reduction of local competition appears to have resulted in an improvement in the condition of surviving adults, and an increase in egg size (Coulson *et al.*, 1982). All of these effects will of course tend to offset the effectiveness of the culling programme.

The low breeding rates of seabirds effectively mean that populations cannot recover quickly from catastrophic declines in the numbers of breeding adults, as might occur as a result of pollution near breeding colonies (see Chapter 6). That seabirds are adapted to maximize adult survival rates means that individual birds will abandon breeding attempts if the perceived risks of death are too great. This is very important in the management of rare species such as the Mediterranean Audouin's Gull, where increased human disturbance is occurring at major breeding colonies. These life-history aspects, and other features of seabird behaviour and ecology outlined in this chapter, must be taken into account if we are to successfully manage and conserve these remarkable birds, as is discussed in more detail in Chapter 8.

CHAPTER THREE

SEABIRD FEEDING ECOLOGY

When not tied to breeding colonies, seabirds can range over vast areas of sea in search of food, and, if necessary, migrate to avoid a hostile winter environment. As the onset of breeding approaches however, they must obtain enough food to undertake the journey to their breeding colonies. During the breeding season itself, in addition to finding food for themselves in the limited areas around colonies, they must find enough food for egg production; when the chicks hatch, they must bring back food of sufficient quality, and at sufficiently short intervals, to support their growing young until fledging. Many seabirds transport food by swallowing it into specially adapted crops or pouches, while others carry food in their bills. Procellariforms have substantially increased their ability to deliver energy to their young by manufacturing stomach oils which are fed directly to the chicks.

Different seabirds have arrived at different solutions to the challenge of obtaining food from the marine environment, and this chapter summarizes how, where and on what seabirds typically feed.

3.1 What seabirds eat

3.1.1 *Range of available prey*

A wide variety of potential vertebrate and invertebrate prey is available to seabirds in the surface waters of oceans and seas, and on the shore and intertidal areas; in addition, several species are able to exploit human waste products from both the fishing industry and, more recently, domestic waste. To a large extent, differences in diet can be attributed to differences in body size, foraging techniques and location (see 3.2 and 3.3); diets can also differ significantly between populations of the same species in nearby colonies due to different foraging opportunities (Schneider and Hunt, 1984).

One tends to associate seabirds with fish, and in sub-arctic areas, where there are large tracts of comparatively shallow so-called 'neritic' waters over the continental shelf, surface shoaling fish are very important to seabirds, especially herrings, sardines and sprats (Clupeidae), sand-eels (Ammodytidae), and capelin (Osmeridae). In tropical areas, small fish are much less available to seabirds by day, coming to the surface only when chased by predatory fish such as tunas; flying-fish and halfbeaks (Exocoetidae) are of major importance as prey.

However, by no means all seabirds are dependent on fish, and a wide range of planktonic crustaceans, cephalopods (mainly squid), molluscs and other invertebrates are also eaten in marine areas. In polar regions, fish are much less important to seabirds, and the availability of large herbivorous crustaceans (particularly the Antarctic krill, *Euphausia superba*) at the sea surface in summer is largely responsible for the enormous abundance of pelagic birds in this area, though carrion and offal are also important in the diet of some species (Ashmole, 1971; Croxall and Prince, 1980).

3.1.2 *Methods of assessing seabird diets*

Quantitative information on seabird diets can be obtained by several different methods, each with different limitations. Indigestible components of food may be vomited as pellets, or evacuated with the faeces, and examination of these can give a good deal of information on the range of items taken. This can be very useful in identifying fish species eaten, since the bony part of the fish inner ear (the otolith) is highly species specific and generally not digested; the size of the otolith also gives an estimate of fish size. However, this works only if the fish's head is eaten, which is not always the case. The drawback of faecal and pellet analyses in studies of seabird diets is that the proportion of indigestible matter is not the same for all types of food, and this method therefore does not indicate the ratio in which different prey species occur in the diet. A further problem arises with the use of this method in assessing the diet of young chicks, since their capacity to digest bony material is often greater than that of adults (Spaans, 1971). The main value of the method is in the study of dietary differences between areas or individuals, or seasonal changes in the diet of birds in a particular locality. Figure 3.1 shows data on the seasonal changes in prey taken by Herring Gulls in a Netherlands colony, based on faecal and pellet examination.

For some seabird species, the food type eaten can be observed directly on the feeding grounds, or inferred from the foraging locality. Direct observation of the food fed to chicks is also possible in species which carry prey in their bills, such as Puffins, or where the parent regurgitates food on to the ground before feeding it to the young, as in the large gulls. However, in the latter case the food may be eaten so quickly by the chicks that the data obtained are biased in favour of the most readily identifiable items. Food regurgitated by adults or chicks on handling provides more information, since it can be examined more closely.

A third method of assessing diet composition involves the examination of stomach contents. Unfortunately, this generally requires the birds to be freshly killed and is therefore unacceptable in many cases, though emetics or stomach pumps can sometimes be used successfully. Differential digestion of material in the stomach can bias the results, and small quantities of hard, indigestible material

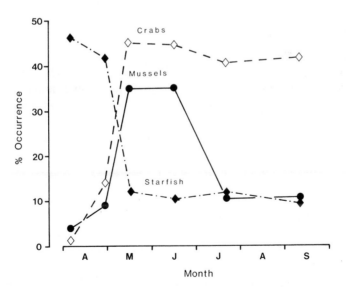

Figure 3.1 Seasonal changes in the extent to which three prey types occurred in the diet of Herring Gulls at a Dutch breeding colony, based on faecal and pellet analysis. (Data from Spaans, 1971.)

may remain in the stomach for some time; the importance of these prey types in the diet may thus be overestimated.

In most cases, a combination of methods is used to obtain information on seabird diets. Since prey types taken often vary considerably between birds and from day to day, an adequate sample is essential, and must be collected over a reasonable time period, otherwise the results obtained may be unreliable (see Bedard, 1976, for a discussion of such problems). The results can then be presented in terms of the number of times a particular type of food was found in samples or, if possible, the proportion of different prey by weight or volume.

3.1.3 *The importance of diet quality*

Obviously, diet quality can have profound effects on survival and reproductive success. Both energetic and nutritive aspects of the food are important, but their significance has been little studied (Montevecchi and Piatt, 1984). Protein quality is generally high in fish, and a number of studies have shown that in gulls, chicks fed on fish grow better than those fed on marine invertebrates (Spaans, 1971; Murphy, 1984). This topic, however, warrants further investigation.

3.2 How seabirds feed

With the exception of specialized deep divers, most seabirds can feed only at or near the sea surface. Food location is predominantly visual, though some nocturnal foragers, such as skimmers, locate prey by touch, and certain procellariforms, such as Fulmars and Sooty Shearwaters, can apparently locate prey by smell (Hutchinson *et al.*, 1984). Flock foraging is very frequent in seabirds, and both visual and auditory cues obtained from feeding birds attract others to the same feeding location (Hoffman *et al.*, 1981).

Seabirds can be broadly divided into those which are good fliers, the aerial types, and those which are good underwater swimmers, the aquatic types. These two specializations are to some extent mutually exclusive, since the long or broad wings necessary for economic flight hinder underwater swimming, for which short narrow wings which can be used as flippers are necessary. Generally the legs of seabirds are not well adapted for walking on land, where they tend to be inefficient and clumsy; however, some, such as gulls, are comparatively unspecialized, being equally at home walking on the shore and flying at sea, and their ability to walk well has allowed them to exploit new feeding opportunities (see 3.3).

A number of different methods by which seabirds obtain their food are described in detail by Ashmole (1971). These are summarized below, and illustrated in Figure 3.2.

3.2.1 *Underwater pursuit diving*

Many seabird species are adapted for swimming underwater, and may pursue their prey to considerable depths. The dive may either be from the water surface, as in penguins, or from the air, as in some shearwaters. Underwater propulsion is obtained in species such as cormorants by using the feet, but most underwater-swimming seabirds use their short wings as flippers. Comparatively little is known about the depth to which these underwater swimmers can dive. Penguins are undoubtedly the most proficient divers, but as a consequence of the extreme wing and body adaptations have lost the power of flight; these birds can dive to great depths, the Emperor Penguin for example reaching over 260 m. Recent studies suggest that auks may also be able to dive to 100 m or more (Piatt and Nettleship, 1985).

3.2.2 *Plunge diving*

Some seabirds which lack specialization for underwater propulsion can nonetheless dive by plunging from the air, thus overcoming their buoyancy, and reach a depth of several metres. Sulids in particular forage by this method, and gulls and terns can also plunge dive, though they penetrate little further than their own body length.

3.2.3 *Feeding from the surface*

Many seabirds feed whilst settled on the water surface, seizing available prey or offal with their bills, a feeding method commonly used by Fulmars, gulls and albatrosses. Species which feed on small planktonic organisms can do this by ducking in water and filtering out the prey, as for instance do Fulmars and Storm Petrels. A specialized filter-feeding technique, known as hydroplaning, is sometimes used by prions; the bird rests its breast on the water surface and, with outstretched wings, propels itself forwards using its feet while keeping its bill submerged.

Figure 3.2 The methods by which different types of seabirds obtain their prey. The figure illustrates underwater pursuit diving using wings (Puffin) and feet (Shag), plunge diving (Gannet), feeding from the surface (Fulmar) and feeding from the air (tern). See text for details.

3.2.4 *Feeding from the air*

Food may be picked up from the water surface if the bird dips down while in flight, as is regularly done by terns and Kittiwakes. Skimmers have a specialized flight foraging technique which involves flying just above the water surface with the bill partially immersed. This is effective only in calm water, but has the advantage that it enables prey to be located by touch and can thus be used in turbid water or at night.

3.2.5 *Shoreline feeding, scavenging and kleptoparasitism*

These are methods by which seabirds obtain food only indirectly from the sea or not at all. Gulls in particular are well adapted to feeding in intertidal areas on a variety of different prey types and on carrion and offal scavenged behind fishing boats, in seaports or on beaches, and indeed even inland (see 3.3).

Feeding by stealing food from other birds (kleptoparasitism) is a way of life for some skuas and frigate-birds, and also occurs to a lesser extent in gulls. The occurrence of kleptoparasitism amongst birds in general appears to be associated with the availability of potential hosts feeding on large, visible food items, and periods of food shortage (Brockman and Barnard, 1979). Large mixed seabird colonies provide ideal opportunities for such behaviour. Frigate-birds are very highly adapted for the agile flying required for kleptoparasitism, which involves chasing other birds until they drop or regurgitate their load; to this end they have reduced the amount of oil in their feathers to such an extent that they can no longer enter the water on which they rely for food (Nelson, 1967). Puffins, terns and Kittiwakes are common victims of skuas in the northern hemisphere, and the skuas selectively parasitize birds carrying bigger and thereby more profitable prey; their success in obtaining the prey is determined by factors such as the duration of the chase, the speed of reaction of the victim and the method of evasion employed (Arnason and Grant, 1978; Furness, 1978).

3.3 Where seabirds feed

The location of seabird foraging areas is governed by a combination of their foraging techniques, breeding areas and prey availability. They concentrate in areas where suitable prey are abundant, such as in many polar and sub-polar regions where intensive water mixing results in nutrient-rich surface waters which support high densities of potential prey (Ashmole, 1971). In some areas, ocean circulation patterns continually bring nutrient-rich water, otherwise confined to the ocean depths, to the surface, as occurs for example in the highly productive upwelling systems off Peru and south-west Africa. The Peruvian system, though subject to occasional catastrophic perturbations (the so-called El Niño—see 5.5.1) is sufficiently stable to support a large and distinctive community of marine animals, including twelve endemic or near endemic species of seabird (Brown, 1980; Schreiber and Schreiber, 1984).

As discussed in section 3.2, seabirds which are adapted for pursuit of prey underwater are comparatively inefficient fliers; such birds cannot glide, and their short wings are suitable only for flapping flight which is energetically very costly. For this reason, almost all of the more specialized diving seabirds (penguins, alcids, diving petrels and diving shearwaters) breed either in polar and sub-polar regions, or upwelling systems, where the considerable abundance of prey makes long foraging trips unnecessary and their foraging technique economical in energetic terms. At low latitudes, where marine prey is comparatively scarce, much longer foraging flights may be required to obtain sufficient food and seabirds must retain the ability to fly long distances without tiring.

The efficiency of seabird foraging also varies in relation to body size. The amount of foraging time required to rear a typical brood decreases with increasing adult body size (Figure 3.3). Thus small species such as terns spend a high percentage of their time finding food for their young, and will be more susceptible to breeding failure during periods of food shortage, since they cannot increase the amount of time spent foraging to compensate (Pearson, 1968; Monaghan and Zonfrillo, 1986).

The fact that gulls are mobile on land, unlike many other seabirds, has enabled them to exploit terrestrial habitats and they are regularly seen feeding in agricultural and urban areas. During winter and periods of hard weather, large numbers forage at refuse tips. Feeding is highly competitive at these sites; the larger (and dominant) male Herring Gulls have higher feeding success than females, and adults of both sexes dominate immature individuals (Monaghan, 1980; Greig et al., 1985). However, the gulls at refuse tips are predominantly females and there is evidence to suggest that the more traditional marine environment is the preferred feeding area (Spaans, 1981; Monaghan *et al.*, 1985).

3.4 Variation in foraging success with age

An improvement in foraging ability with age has been found in virtually every seabird in which adult and young birds have been compared (Burger, 1980). Most such studies have been carried out on gulls and terns, which are comparatively easy to observe. In a wide variety of different foraging situations, young gulls have been shown to be less efficient than adults, and are often forced into sub-optimal areas due to competition from older birds (Monaghan, 1980; Burger, 1980).

The improvement in foraging performance in gulls has been shown to continue over several years (Figure 3.4) and presumably the

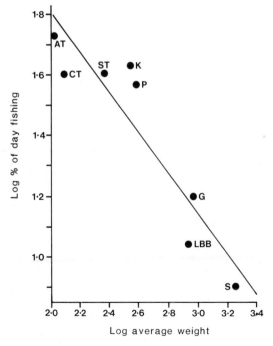

Figure 3.3 The relationship between the proportion of the day spent finding food to rear a typical brood and adult body weight. The smaller the adult size, the greater the amount of time spent foraging. (Data from Pearson, 1968.) AT = Arctic Tern; CT = Common Tern; ST = Sandwich Tern; K = Kittiwake; P = Puffin; LBB = Lesser Black-backed Gull; S = Shag; G = Guillemot.

lack of adequate foraging skills contributes to the delayed maturity characteristic of these birds (Greig *et al.*, 1983). Young breeders also tend to have lower reproductive success than adults (see 2.5.2) and, while this may in part be due to lack of experience of the mechanics of breeding itself, foraging skills may continue to improve even after breeding has commenced. Ainley and Schlatter (1972) examined the fledging weights of chicks reared by Adelie Penguins of different ages. Older birds fledged heavier chicks and bigger broods than did younger birds (Figure 3.5). This effect did not appear to be related to breeding experience, which suggests that as the bird gets older, its ability to feed chicks depends on what is learned from accumulated experience at sea, rather than breeding *per se*.

3.5 Studying seabird communities

Seabird breeding communities have been studied in a variety of areas, for example in the Barents Sea (Belopoliskii, 1961), the Pacific

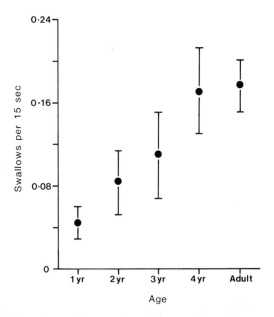

Figure 3.4 The relationship between foraging success and age in Herring Gulls feeding at a refuse tip. Foraging performance continues to improve until at least the 4th year of life. (Data from Greig *et al.*, 1983.)

Ocean (Bedard, 1969; Ashmole and Ashmole, 1967) and the North
Sea (Pearson, 1968). Such studies attempt to evaluate the energetic
demands of these communities (Chapter 5), and to determine the
level of interspecific competition.

In general, some degree of ecological segregation of foraging areas
has been demonstrated for species breeding in the same areas, in
terms of prey types taken, foraging ranges or foraging depths. For
example Ashmole (1968) studied the foraging patterns of five species of terns breeding on Christmas Island in the Pacific Ocean. All
five species fed mainly in flight, catching fish and squid chased to
the surface by predatory tuna. The similarly-sized Sooty Tern and
Brown Noddy were found to catch prey of similar sizes, but the
Sooty Tern foraged much further offshore. The smaller Fairy Tern
and Black Noddy also ate similar-sized prey, but concentrated on
different prey types and foraged at different times of day. The smallest
of the five species, the Blue-grey Noddy, was found to eat mainly
fish larvae and very small invertebrates. Ashmole found a general
relationship between body size and the size of prey eaten, as did

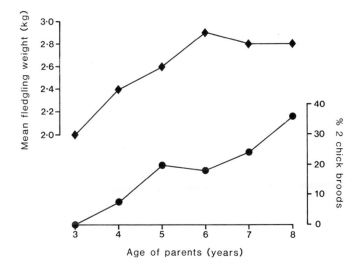

Figure 3.5 The relationship between fledging weight of chicks and parental age
(upper graph), and proportion of two chick broods successfully reared and parental
age (lower graph) in Adelie Penguins. Older parents fledge heavier young and are
more likely to rear two chick broods. Breeding performance continues to improve
until at least the 8th year of life. (Data from Ainley and Schlatter, 1972.)

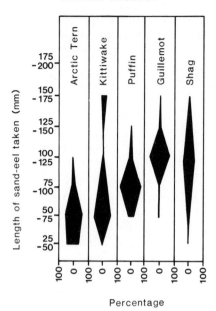

Figure 3.6 The sizes of sand-eels taken by seabirds of different sizes. Bigger birds tend to feed on bigger prey. (Data from Pearson, 1968.)

Pearson (1968) in his study of a greater diversity of seabirds in a North Sea colony (Figure 3.6). Pearson also found clear differences in foraging ranges and foraging depths between the ten species examined, which further reduced the level of interspecific competition.

Non-breeding associations of seabirds have been little studied, largely due to the difficulties in obtaining sufficient information on their foraging behaviour (Baltz and Morejohn, 1977; Ainley and Sanger, 1979).

CHAPTER FOUR

REGULATION OF SEABIRD POPULATIONS

4.1 Introduction

For any animal population in a defined area the change in population size from one year to the next is given by the equation

$$\text{Population change} = (\text{Recruitment} + \text{Immigration}) - (\text{Mortality} + \text{Emigration}).$$

One or more of these four factors must play a key role in the regulation of seabird populations. It is particularly important to understand this relationship, since conservation of endangered species, control of pest species, sensible interpretation of many monitoring studies and simulation models depend on a knowledge of how seabird populations are regulated.

In this chapter we shall look first at the various theories put forward to explain how seabird population sizes are controlled, and then consider how recent evidence supports or contradicts these ideas.

4.2 Regulation of populations: the theories

Seabirds differ from most other groups of birds in a number of aspects of population dynamics and ecology. In particular, they tend to have small clutch sizes, deferred maturity and low adult mortality rates (Chapter 2). Almost all species are colonial breeders, and forage at a variable distance from the colony on marine animals that may be available in a temporally and spatially unpredictable manner (Chapter 3). An unpredictable, and in some species also nutritionally poor, food selects for slow chick growth (see 5.2). These factors together result in low reproductive output which is combined with a high life expectancy of adults, characteristics of K-selected species (see 2.4).

The fact that seabirds are towards the K end of the r–K continuum, and so should be selected for competitive ability rather than for fecundity, should lead us to expect that social and other biotic factors will figure prominently in their population dynamics, while environmental factors may be of less importance. For this reason we should not be too surprised if we find that seabird populations are not controlled in the same way as populations of insects, or even small passerines. For reviews of the controversies over the methods of regulation of animal and bird populations see Krebs (1978) and Perrins and Birkhead (1983). The arguments hinge on the roles of density-dependent versus density-independent factors and on the relative importance of different density-dependent mechanisms.

4.2.1 *Lack's hypothesis*

Starting from the observation that bird populations remain more or less stable in numbers, Lack argued that they are controlled around an equilibrium value by density-dependent factors. In other words, a high population density leads to a lowering of the density. Lack argued that birds lay the number of eggs that result in the largest possible number of surviving offspring, and stated that, since reproductive output shows little correlation with population density, the regulation of population size must come about as a consequence of density-dependent mortality. Since food supplies are least in winter, and in many species winter mortality is high, Lack argued that starvation in winter due to density-dependent competition for food was the key factor regulating bird numbers. Considering seabirds in particular (Lack, 1967), he saw the possibility that competition for food close to large breeding colonies might influence clutch size and age of maturity, and that a lack of sites suitable for breeding colonies might limit numbers in a few areas. However, he considered that these factors would be trivial compared to the effect that high numbers following breeding had in increasing competition for food during winter, so that this would also be the mechanism regulating population sizes of seabirds.

Lack (1966) stated that even if breeding sites were limited in numbers, and so prevented some birds from breeding, total numbers in the population would continue to increase (i.e. the proportion of non-breeders would rise) until eventually controlled by food. Although this may sometimes occur, Lack's view is not necessarily correct, as is shown by the following worked example.

Model seabird population:

Number of available nest sites	100
Average annual survival of adults	0.80
Average number of young fledged per pair per year	2.0
Average survival to breeding age	0.25

If population = 200 adults:
Annual loss of adults = 200 × 0.20 = 40 die
Annual fledgling production = 100 × 2.0 = 200 fledglings
Number recruiting as breeders = 200 × 0.25 = 50 recruits
Population change = +50 −40 = +10 (i.e. a rate of increase of +5%).

If population = 300 adults:
Annual loss of adults = 300 × 0.20 = 60 die
Recruits remain the same since 100 pairs breed = 50 recruits
Population change = +50 −60 = −10 (i.e. a rate of increase of −5%)

If population = 250 adults:
Annual loss = 250 × 0.20 = 50 die
Recruitment = 50 (as before)
Population change = 0

So if food supplies could support 300 adults (plus their young) then the population would never become limited by food since the limited number of breeding sites would result in an equilibrium population of 250 adults. If food supplies could support only 220 adults (plus their young) then the population would reach an equilibrium of 220 adults, of which 200 would breed and 20 would be non-breeders.

We can conclude from this worked example that availability of nest sites could in theory limit population size in certain circumstances, although no-one has suggested that it normally does so.

4.2.2 Ashmole's hypothesis

Ashmole (1963) agreed with Lack that seabird numbers are regulated by density-dependent processes related to food availability, but considered that the effects of food supplies would be seen during the breeding season rather than during the winter, at least in the case of tropical seabirds. He felt that in tropical regions the food availability will change little between the breeding season and non-breeding periods, and while seabirds are able to disperse over wide areas of ocean outside the breeding season, when nesting they are confined to foraging within a certain distance of the colony. They also require a higher food gathering rate in order to raise young. Thus competition for food should be most severe during the breeding season. Combined with this, life-history theory would predict that seabirds should sacrifice their young in order to protect their own survival, since long-lived seabirds can attempt to breed many times but each attempt has a low likelihood of contributing to future generations (see 2.5). The result of the competition envisaged by Ashmole should be a density-dependent reduction in breeding performance of tropical seabirds.

Consider a small seabird colony on a remote island. Each pair would find plenty of food in the sea close to the colony, so would not have to work hard to raise its chicks. With high chick production, colony size would increase. The larger numbers of breeding birds will now begin to deplete their food supplies close to the colony, so will have to fly further out to sea to get food. Longer flights require more time at sea, and more food to fuel the adults' activity. Eventually, as numbers continue to rise, birds will be travelling so far that they will not have time to make as many flights each day as are needed to keep their chick growing. Breeding success will fall and result in density-dependent regulation of the size of the colony. These processes are illustrated in Figure 4.1.

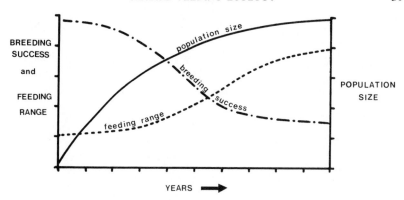

Figure 4.1 Relationships between colony size, foraging range, and breeding success as hypothesized by Ashmole's model of seabird population regulation.

Of course the situation in high latitudes, where there is a seasonal flush of marine production in the summer, may be different, and Ashmole left open the question of whether his model applies only to tropical seabirds or to all.

It is difficult to imagine that seabirds come into active competition for food when thinly spread over the oceans in winter, as required by Lack's model, yet competition is easy to imagine in the immediate vicinity of a large seabird colony, a further attractive feature of Ashmole's model. Some recently published estimates of seabird population densities around breeding colonies and in non-breeding areas are shown in Table 4.1. The densities found during the summer are so much greater than those in winter that it seems likely that Ashmole's model is as plausible as Lack's. We shall return to consider the evidence for and against each model later in this chapter.

Table 4.1 Densities of seabirds around colonies during the breeding season and in the wintering range during winter.

Species	Colony region	Numbers per km^2	Winter region	Numbers per km^2	Source
Gannet	<37 km from Noss	1.5	North Sea	0.04	Tasker et al., 1985
Gannet	Inshore Bass Rock	1.0	North Sea	0.04	Tasker et al., 1985
Razorbill	Orkney and Shetland	0.3	North Sea	0.05	Blake et al., 1985
Puffin	Orkney and Shetland	1.7	North Sea	0.05	Blake et al., 1985
Guillemot	Fair Isle	30	North Sea	1.6	Blake et al., 1985
All seabirds	<60 km Pribilof	129	E. Bering Sea	<10	Hunt, in press.

4.2.3 *Wynne-Edwards' hypothesis*

Wynne-Edwards (1955, 1962) also considered that density-dependent processes regulate seabird populations. He argued that starvation is rarely found in bird populations; food shortage in winter does not seem to lead to mass mortality as hypothesized by Lack. Wynne-Edwards suggested that birds regulate their numbers to a level determined by social interactions and set below the limit that would be imposed by food supplies. He argued that seabirds, with their small clutch sizes, deferred maturity, non-breeding adults and high adult survival rates, provide an extreme example of prudential restraint to avoid overpopulation.

Regulation of numbers by social convention requires group selection in order to operate, since natural selection will favour the most productive members of the group. In order for Wynne-Edwards' model to operate, groups that overproduce must become extinct while groups that show restraint survive. Theoretical studies of group selection suggest that the necessary conditions are rare; groups need to be small and discrete, with very little gene-flow. Seabird populations tend to be large, with much emigration from natal colonies to others. This has led ecologists to consider regulation of seabird populations by prudential restraint to be very unlikely.

4.2.4 *Control by density-independent processes*

Andrewartha and Birch (1954) attacked the concept of density-dependent regulation. To some extent all mortality factors may be density-dependent to some greater or lesser degree. For example, weather is generally considered to be a density-independent mortality factor, but where the number of sheltered roost sites is restricted, mortality caused by bad weather will be density-dependent. However, density-dependent factors may not necessarily regulate a population even if they do exist. Where density-independent mortality predominates, a population may never increase to the size where density-dependent processes become important. Andrewartha and Birch argued that animal populations are controlled by varying mortality and natality rates in different patches within the environment. Numbers continually fluctuate in these patches, with migration between patches counteracting local extinctions. This density-independent model works because the dynamics of the population varies from patch to patch. It seems to be quite suitable for describing the dynamics of many insect species, for which habitat heterogeneity is clearly important, but most ecologists consider that density-dependent regulation is more important in vertebrates, which tend to be less affected by environmental factors such as weather, and more influenced by biotic factors, such as competition for food and mates.

4.3 The evidence: are seabird populations regulated?

Most seabird biologists would probably answer that they are. Demonstrating density-dependent regulation is another matter. Furthermore, there is clear evidence that many seabird populations have changed dramatically in numbers with little sign of an equilibrium size around which they are held.

Some population declines in the past were the result of excessive exploitation, and many recent population increases may be due to subsequent protection. Overfishing by man can result in both increases and decreases in seabird numbers, depending on the trophic relationships between the seabirds and the exploited fish (Furness, 1982). It has often been suggested that the increases in seabird numbers around the British Isles since the end of last century were due to cessation of exploitation by man. While this may partly be true, we know that approximately 50% of the Fulmar chicks at St

Table 4.2 Rates of increase of Fulmar populations on St Kilda and the rest of Britain and Ireland. Data from Fisher (1966) and Cramp et al. (1974)

| Years | Number of occupied sites at end of decade | | Mean annual rate of increase | | |
	St Kilda	others	St Kilda	Others	Three-year running mean
1879	25 000	24			
1879–1889	25 000	101	0%	17%	
1889–1899	25 000	478	0%	18%	17%
1899–1909	25 000	2039	0%	16%	15%
1909–1919	25 000	5969	0%	11%	12%
1919–1929	25 500	13 482	0%	8%	10%
1929–1939	20 800	35 223	−1%	10%	8%
1939–1949	38 200	70 472	6%	7%	7%
1949–1959	37 500	97 039	−0.2%	3%	7%
1959–1969	36 600	269 039	−0.2%	10%	

Kilda in the Outer Hebrides were harvested annually over a period of several centuries, with no detrimental effect on the breeding population size, but since the evacuation of the human community in 1930, no Fulmar chicks have been harvested, so that breeding success of the colony probably doubled at that time. Despite this, numbers of Fulmars breeding at St Kilda have increased only very slightly, if at all, suggesting that harvesting of chicks had not held the population in check. In contrast, numbers of Fulmars elsewhere in Britain have shown a dramatic increase after colonization of Shetland in the 1870s from the Arctic (Cramp et al., 1974). The pattern of spread of Fulmars after this colonization suggests that it took place from the north and that St Kilda birds were not involved. Diet and feeding behaviour of Fulmars at St Kilda and in Shetland are quite different (Furness and Todd, 1984) and this suggests that new feeding conditions around Shetland may have encouraged the spread of Fulmars. However, there is little sign that the doubling of chick production on St Kilda after 1930 had any effect on the breeding numbers there. The increase of the Fulmar population in other parts of Britain does give slight support for the idea that numbers are regulated in a density-dependent manner, since the rate of increase of the Fulmar population has fallen over the decades in a way that is compatible with the progressively increasing effect of a check to population growth (Table 4.2).

The number of Kittiwake colonies in England, Wales and the Isle of Man increased from 18 in 1900 to over 62 in 1969, but increased by only one between 1969 and 1979 (Figure 4.2). Population trends differed between regions and between consecutive decades (Table 4.3). The prolonged increase to 1969 may be partly due to increases in food supplies, but it is at least partly a recovery from human persecution during the 19th century. Density-dependent regulation has not played a significant part in Kittiwake population dynamics over that period. Coulson (1983) speculates that the cause of the declines seen in some regions since 1969 is probably food shortage, since no other factors appear to be implicated. He points out that food shortage when Kittiwakes are dispersed in winter is unlikely to be operating, since that would lead to population declines of Kittiwakes from all regions. Food shortage affecting birds at breeding colonies appears the likely mechanism.

The world population of Gannets has increased from 67 000 pairs in 1909 to 82 500 pairs in 1939, 196 500 pairs in 1969 and

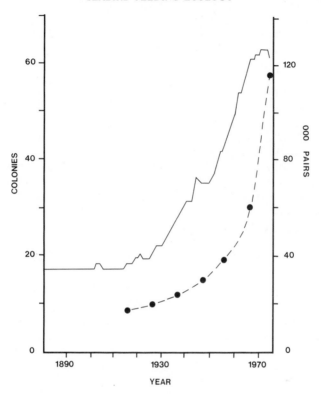

Figure 4.2 The numbers of pairs and numbers of colonies of Kittiwakes in England, Wales and the Isle of Man this century. Continuous line, colonies; broken line, breeding pairs. (From Coulson, 1983.)

213 000 pairs in 1976. As with Kittiwakes, persecution of Gannets undoubtedly reduced numbers last century, but they now breed in colonies formed in many new places and probably in larger numbers than ever before, with no clear sign of density-dependent checks operating so far. Either their numbers are not regulated, or the population has still not reached a limit set by food. The latter is quite possible, since changes in North Sea and North Atlantic fish stocks this century have been favourable to the Gannet: there are more smaller fish on which it can feed (section 5.8).

We can also examine changes in seabird numbers using the fact that guano harvests in Peru and South Africa correlate closely with the sizes of seabird populations in those areas (Nelson, 1978;

Table 4.3 Changes in numbers of breeding pairs of Kittiwakes in eight regions of the British Isles between 1959–69 and 1969–79. Data from Coulson (1983).

Region	Percentage population increase over decade	
	1959–1969	1969–1979
East Scotland	+78%	+37%
East England	+42%	+81%
South England	+130%	−16%
South Wales	+59%	−3%
North Wales and NW England	+65%	−20%
West Scotland (excluding Ailsa Craig and St Kilda)	+10%	−15%
Ailsa Craig	+4%	−80%
St Kilda	+50%	−61%
East Ireland	+343%	+32%
South Ireland	+186%	−56%

Crawford and Shelton, 1980). It is possible to show correlations between seabird breeding populations and the size of fish stocks in these areas (see for example Figures 5.11 and 5.12). This suggests regulation of breeding numbers by food availability. A particularly long time series for Peruvian seabirds (Nelson, 1978; Santander, 1980) shows that numbers have fluctuated over a wide range, with no tendency to remain around an equilibrium (Figure 4.3). Since 1953 the numbers of seabirds have been influenced by the fishery for anchovies and their decline can be related to a reduction in the anchovy stocks. Earlier increases in seabird numbers have been ascribed to protection from human disturbance and to an increase in the area made available for nesting by guano managers (Duffy,

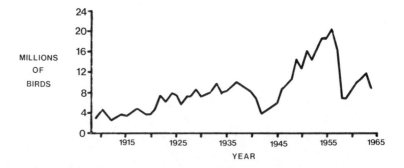

Figure 4.3 Fluctuations in numbers of guano birds breeding on islands off Peru as estimated from guano harvests 1908–1964. (From Santander, 1980).

1983), while periodic fluctuations are caused by the oceanographic changes resulting from El Niño (see 5.5.1). If correct, this would mean that over the last 100 years Peruvian seabird numbers were firstly unregulated, then regulated by nest site availability, and now are regulated by food shortage.

We can conclude that historical records of seabird population sizes show large fluctuations in some and systematic trends over considerable periods in others. While some of these can be correlated with changes in food supplies, the historical record provides little evidence for density-dependent regulation of numbers.

4.4 The evidence: density-dependent effects

Even in remote polar regions seabird populations are influenced by man (May et al., 1979; Evans and Waterston, 1976) so it is difficult to observe seabirds devoid of man's influence. This confounds the problem of determining how their numbers are regulated. Their long lifespan also precludes most experimental approaches. However, a number of studies have investigated the influence of population density on aspects of population dynamics.

4.4.1 Limitation by nest sites

There are far fewer seabird colonies than marine productivity of some regions would lead one to expect. Brown (1979) highlighted the upwelling region off Senegal where at very few sites could seabirds nest safe from terrestrial predators. Provision of artificial islands off South-west Africa and Peru resulted in seabird population increases there (Crawford and Shelton, 1978; Duffy, 1983). Limitation by availability of colony sites is rarely a problem in Arctic, northern temperate or sub-Antarctic areas; many authors point to an abundance of available unused or incompletely used sites.

Within colonies the number of nest sites may be a limiting factor. The consequence should be seen as:

(i) Competition for sites
(ii) A reduction in mean breeding success as more poor-quality nest sites become occupied
(iii) Increasing numbers of floating non-breeders unable to obtain sites
(iv) Increasing emigration.

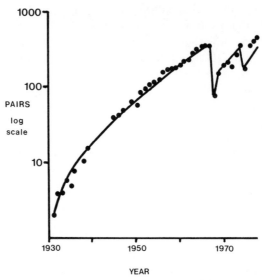

Figure 4.4 Growth of the Shag population breeding on the Farne Islands since 1931; output from a simulation model represented by dotted line; dots are actual nest counts (logarithmic scale). (From Potts *et al.*, 1980.)

Many seabirds return to breeding colonies many months before the onset of breeding, possibly in order to defend nest sites against competitors. Where populations have been rapidly increasing there is often a tendency for birds to reoccupy sites progressively earlier in the year (Coulson and Wooller, 1976; Taylor and Reid, 1981) suggesting an increase in competition for sites.

Potts *et al.* (1980) ranked Shag nest sites on the Farne Islands according to four measures of quality: protection from high seas, exposure to rain, capacity to hold large chicks, and access to the sea. They showed that breeding success was lower for pairs with low-quality sites and that average nest-site quality declined as breeding numbers increased. They were able to construct a simulation model of the growth of the Shag population which closely fitted actual numbers (Figure 4.4) using only the known annual survival values and the measured relationships between nest-site quality and breeding success. Discrepancy between the model and reality since 1975 was attributed to a change in immigration or emigration rates.

This important study shows that the Farne Island Shag population suffers highly variable adult mortality from year to year, but numbers at this colony are limited by a density-dependent influence of nest-site quality on breeding success.

In contrast, the long-term study of Fulmars at Eynhallow, Orkney, showed that breeding numbers, which increased to a peak in 1970 and then declined, could best be simulated in a model using only adult survival and recruitment as inputs (Ollason and Dunnet, 1983). Breeding success could not be shown to be related to any measured characteristic of the nest sites (Olsthoorn, 1984) and over 700 nest sites were used between 1950 and 1977 by a population of no more than 241 breeding pairs in any one year (Dunnet *et al.*, 1979). Clearly Eynhallow Fulmar numbers have not been limited by nest sites.

The existence of a floating non-breeding surplus is difficult to demonstrate. Removal experiments have shown the existence of floaters in Cassin's Auklet (Manuwal, 1972). Similarly, density-dependent effects on recruitment and immigration or emigration rates are difficult to study. By comparing rates before and after a cull that reduced breeding numbers of Herring Gulls to a quarter of their earlier numbers, Coulson *et al.* (1982) showed that a smaller proportion of young emigrated at low population density, suggesting density-dependent competition for nesting territories. Birkhead and Furness (1985) concluded, in a review of population regulation in seabirds, that limitation by nest sites appears to be an exceptional rather than a general situation.

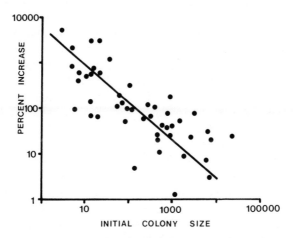

Figure 4.5 The relationship between Kittiwake colony size and proportional increase between 1959 and 1969. (From Coulson, 1983.)

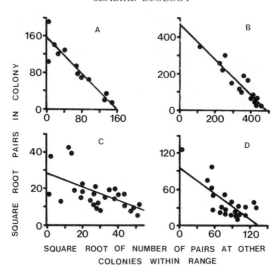

Figure 4.6 Relationships between sizes of colonies and numbers at other colonies within a specific shortest sea distance for (*a*) Gannets (100 km), (*b*) Puffins (150 km), (*c*) Shags (30 km), and (*d*) Kittiwakes (40 km). (From Furness and Birkhead, 1984.)

4.4.2 *Colony sizes and growth*

Small Kittiwake colonies in Britain showed a greater rate of increase than larger ones between 1959 and 1969 (Figure 4.5). Coulson (1983) suggested that this was because smaller colonies are more attractive to potential recruits. This could be because there is less competition at smaller colonies for sites, or for food around the colony.

Colony sizes also appear to be influenced by numbers of conspecifics at neighbouring colonies. Gannets, Puffins, Kittiwakes and Shags breed in larger numbers in colonies that are not surrounded by other large colonies whose members may compete for local food supplies (Furness and Birkhead, 1984). As a consequence, there is a tendency for colony size to be negatively correlated with numbers of nearby conspecifics (Figure 4.6). Similarly, colonies on small offshore islands in Alaska are, on average, larger than colonies on promontories, which are, on average, larger than those on linear coastlines (Figure 4.7) (Birkhead and Furness, 1985). This would be predicted if numbers are related to food availability near to the colony, since birds on offshore islands can forage in an arc of 360° of sea, twice as much as is available to birds on linear coastlines.

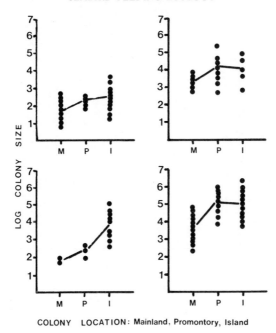

COLONY LOCATION: Mainland, Promontory, Island

Figure 4.7 Colony size (logarithmic scale) in relation to colony type (classified as mainland, promontory or island) for Alaskan seabird colonies. (From Birkhead and Furness, 1985.)

These patterns support, but do not prove, Ashmole's model of population regulation by food supplies near to the colony.

4.4.3 *Feeding distribution of colonial seabirds*

Tropical seabirds feed at the surface of the sea, so that the available prey is a function of feeding range. If prey abundance determines population size of a predator, pelagic feeders should be more abundant than inshore feeders. Analysis of seabird communities of nine tropical oceanic islands supports this prediction (Diamond, 1978). Also, species that migrate to winter elsewhere were more numerous than residents, which Diamond took to indicate that limiting mortality took place outside the breeding season.

Considering seabirds foraging from a colony during the breeding season, it is clear that the foraging density does decrease with distance from the colony as required by Ashmole's model, as demonstrated

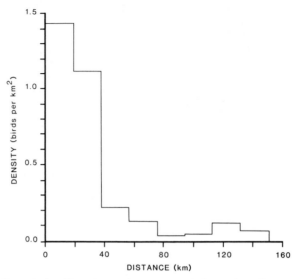

Figure 4.8 Gannet densities at sea in summer at increasing distances from the colony at Noss, Shetland. (From Tasker *et al.*, 1985.)

for example for Gannets by Tasker *et al.* (1985) (Figure 4.8). Similar distribution patterns have been shown for Common Guillemots around Fair Isle, where breeding adult density fell rapidly to very low levels beyond 12 km and nonbreeders foraged at generally greater distances than breeders (Blake *et al.*, 1984), and for Common Guillemots, Puffins, Razorbills, Fulmars and Great Skuas from Foula (Furness, 1983). These patterns imply that birds can reduce intraspecific competition for food, if this occurs, by travelling further from the colony to feed. The cost of travelling further is that this takes time which could be devoted to other breeding activities, and involves energy costs of flight, which are high in species that use flapping flight.

4.4.4 *Breeding performance in relation to colony size*

The growth rate of Brunnich's Guillemot chicks varies between years and among different areas of a single colony. However, these differences are small compared to consistent differences between colonies, and there is a negative correlation between colony size and chick fledging weights (Figure 4.9). The cause of this was interpreted by

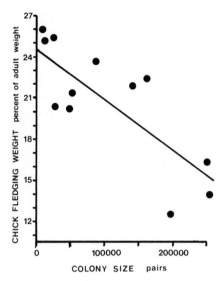

Figure 4.9 Brunnich's Guillemot chick fledging weights, as a percentage of adult weight, in relation to colony size. (From Gaston *et al.*, 1983, with additional data from Furness and Barrett, 1985.)

Gaston *et al.* (1983) to be that birds from larger colonies had to travel further to find food, so bringing less food back to the chick. This is exactly the mechanism proposed by Ashmole for regulation of tropical seabird populations.

Culling of Herring Gulls on the Isle of May has resulted in an increase in egg size, reduced age of first breeding, and an increase

Table 4.4 Relationships between seabird population size and breeding performance (* $p < 0.05$, ** $p < 0.01$). From Birkhead and Furness (1985).

Species	Reproductive parameter	Correlation with log colony size
Kittiwake	Clutch size	−0.67 **
	Chick growth rate	−0.40
	Breeding success	−0.68 **
Common Guillemot	Fledging weight	−0.37
	Chick growth rate	−0.81
	Breeding success	−0.43
Brunnich's Guillemot	Fledging weight	−0.57 *
	Growth rate	−0.52
	Breeding success	−0.27

in both size (wing length) and weight of adult birds remaining (Coulson *et al.*, 1982). These density-dependent effects may all be due to an increased supply of food per bird since the culling reduced numbers.

Comparisons of Kittiwake breeding performance at colonies throughout the species' range also show a tendency for breeding success to be highest in the smallest colonies (Table 4.4). While numbers of conspecifics influenced performance, numbers of other species of seabirds did not (Birkhead and Furness, 1985), suggesting that interspecific competition was of little significance in determining population sizes.

4.5 Conclusions

There is still a need for further studies of the effects of colony size on performance before we can conclude that this is a common or universal phenomenon, and it has yet to be shown that these effects regulate numbers as proposed by Ashmole. A recent review of this subject (Birkhead and Furness, 1985) concluded that colony sites may in some rare cases limit numbers. In other rare cases, breeding sites within colonies may limit local numbers, and density-dependent competition for food during the breeding season may be an important mechanism. Food shortage in winter may influence the ability of an individual to compete for food during the breeding season, so the discrete models proposed by Lack and Ashmole may in fact both be partly correct. Further detailed tests of these hypotheses are required before the issue can be considered settled. In particular, the possibility of prey depletion around seabird colonies needs to be tested by observation or experiment. It seems clear that sedentary fish populations around colonies might become depleted, and some preliminary studies of fish densities around cormorant colonies by seabird ecologists in Newfoundland suggest that this does in fact happen. However, many seabirds feed on fish stocks that migrate (e.g. capelin, herring, mackerel), and it is difficult to see how Ashmole's model would operate where fish numbers are continuously replaced by migration into the vicinity of a seabird colony. Many challenging areas of seabird population regulation clearly remain to be elucidated.

CHAPTER FIVE

INTERACTIONS WITH FISHERIES

5.1 Introduction

The oceans cover 70% of the earth's surface and provide man with an annual harvest of around 60 million tonnes of fish (1 tonne = 1000 kg). The worldwide harvest has increased over the years as more and more fish species become commercially worth exploiting and fishing techniques develop. However, many fish stocks have been heavily exploited and yields from some of these have fallen. In some cases these declines can be attributed to overfishing, in others to climatic fluctuations, but declining commercial catches have sometimes led to suggestions from fishermen that natural predators such as seals and seabirds should be culled in order to reduce their impact on stocks. Where it is believed that this competition may result in serious and avoidable financial loss to fishermen, an assessment of economic impact of seabirds and seals is a prerequisite to sensible management.

Seabirds may interact with commercial fisheries in a number of ways, the most important of which are the following.

(a) *Seabirds as competitors for fish.* Seabirds might reduce commercial catches by eating large numbers of fish that would otherwise be caught by man. In order to determine whether this happens we need to know the sizes and distributions of seabird populations, the quantity of food consumed, composition of the diets in terms of species and sizes, the production of the fish stocks and the proportions of the stocks taken by the fishing industry.

(b) *Waste from fishing vessels.* Some fisheries discard large quantities of edible waste into the sea, and this is fed on by scavenging seabirds. It has been argued that this has made available to certain seabirds food supplies that they could not otherwise exploit, and so has led to dramatic increases in numbers and breeding ranges of some species. The most commonly cited examples of this are the Fulmar in the North Atlantic and the large gulls in many parts of the world.

(c) *Changes in ecosystem structure.* Generally, commercial fisheries exploit first the larger species of fish which feed at the highest trophic levels, since these tend to be the most profitable fish to catch. Heavy exploitation of stocks of large predatory fish results in a decrease in the average size of fish in the population. Stock biomass may also fall. These changes can have important consequences where seabirds feed on the same prey as the exploited predatory fish. Heavy exploitation of large predatory fish may mean that more small prey fish will be available for seabirds, allowing their numbers to increase. Other changes in energy flow within marine food webs may be detrimental to seabirds. If seabirds feed on a fish stock that is heavily exploited by man, then a reduction in stock size may affect seabirds before it causes the fishery to reduce its catching effort.

Later in this chapter we shall review some of the important changes in marine ecosystems which have taken place and the interactions which have resulted between fisheries and seabird populations. First we shall consider how we can estimate the impact of seabirds on fish stocks. To do this we start by estimating how much food seabirds eat.

5.2 Estimating food consumption by seabirds

Because of the environment in which seabirds live and their wide-ranging behaviour, it is extremely difficult or impossible to obtain direct measurements of the amount of food they eat. As a consequence, various indirect methods have been developed to estimate this amount. Broadly speaking, these can be separated into three groups of increasing precision, difficulty and cost:

(i) Observations of seabirds at breeding colonies to determine sizes of food loads brought to the colony and the number of feeding trips made per day

(ii) Combination of time–activity budgets of seabirds observed in the field with measured metabolic rates of the birds under controlled laboratory conditions

(iii) Direct measurement of metabolic rates of free-living seabirds

We shall now consider each of these methods in turn.

5.2.1 *Field measurements of feeds and feed weights*

In many seabird communities, particularly at high latitudes, one or two species are numerically dominant and are responsible for almost all the food consumption of the community. Field observations during the breeding season can be used to determine how many feeding trips the average adult makes per day and a sample of adults returning to the colony with food can be collected (by shooting or live-catching) to discover how much food is carried in an average meal. The number of adults at the colony can then be multiplied by the number of feeding trips and the average weight of a meal to give the daily food consumption of the population, and this can be multiplied by the number of days that the birds spend at the colony each summer to give the annual food consumption of the adults in the area.

Swartz (1966) used this method to assess the quantity of food consumed by seabirds at Cape Thompson, Alaska (where guillemots predominate), and obtained a total of 13 100 tonnes of fish per year taken by the 421 000 adult guillemots. This is a large amount, but we do not know how far these guillemots were travelling to find food so we cannot place it in the context of fish production. More importantly, we have no measure of the accuracy of this calculation. It is not easy to measure load sizes in seabirds, since part of any meal will have been digested before the bird arrives back at the colony, and birds often regurgitate when caught or shot. Some arbitrary decision must be made as to when a bird is 'empty' and when it contains a small load, since small amounts of food may remain in the stomach for some time after earlier meals. Counting the number of feeding trips per day is also difficult. Some individuals making exceptionally long trips may in fact be digesting one load at sea before catching another and bringing it to the colony. Some trips to sea may not be for feeding. Again, some rather arbitrary decisions must be made as to how long a bird must be away from the colony for that departure to be classified as a feeding trip.

Several studies have been carried out to measure feeding rates of colonial seabirds. As an example of the inaccuracies that may arise in these, consider studies of guillemot colonies. Estimates of daily consumption by adult guillemots vary from 30 g per day (Kaftanovski, 1951) to 220 g per day (Tuck and Squires, 1955), with half-a-dozen other estimates by other workers falling between these extremes. Some of the difference may be due to variations in climate, food availability, adult activity budgets or the calorific value of the food taken, but we can conclude that these simple direct estimates of food consumption by seabird populations at colonies are based on rather inaccurate data and so can only be used to provide very approximate estimates of the amount of fish taken by seabirds. A further problem is that they do not include nonbreeding birds associated with the colony, nor can they be used for studies outside the breeding season or away from colonies.

5.2.2 Bioenergetics equations

If we know how much time birds spend in each activity and the energetic cost of each activity, then we can estimate their overall energy requirements. This 'integrative bioenergetics' approach gives us an alternative indirect method which can be based on firmer and testable foundations. The use of bioenergetics methods to assess food requirements of bird populations is reviewed by Kendeigh et al. (1977), and its application to seabird populations by Furness (1982) and Wiens (1984). In this section we will be concerned with how we can measure the energy requirements of birds under a variety of controlled conditions.

Metabolic rates can be measured in the laboratory by analysis of gaseous exchange; consumption of oxygen or production of carbon dioxide. This allows very precise and accurate determinations of energy requirements, which can be translated to food consumption if we also have figures for the calorific value of the food and the digestive efficiency (proportion of calories in the food eaten that are assimilated and so available for metabolism). Alternatively, we can measure the amount of food actually consumed over a period of days. Providing the bird under investigation does not deposit energy stores (e.g. fat) or deplete its energy stores over this period, then the quantity of food consumed gives a direct measure of the food requirements under the conditions of that period. This latter approach is less satisfactory in a number of respects. It is difficult to ensure that captive birds do not change in tissue composition.

A stable weight could be due to a loss of water equal to the gain in fat, for example. It is also much more difficult to maintain constant environmental conditions over a period of days than for the minutes required for studies of respiratory exchange. However, certain costs cannot be measured so easily by gaseous exchange methods. Moulting and growth are two clear examples, where the energy costs of these activities are spread over a long time period and add to other costs of living. In these cases, longer experiments where food intake is monitored can provide the best measure. For most activities, determinations by measurement of oxygen consumption are better. Metabolic rates are measured in terms of energy, expressed as kilojoules (kJ) or kilocalories (kcals), where 4.184 kJ = 1 kcal.

5.2.3 *Basal Metabolic Rate (BMR)*

The metabolic rate of a seabird depends on temperature, photoperiod, body weight, plumage condition, time of year, activity, exposure to wind and sun, whether the bird is growing new feathers, and whether it is digesting food. In order to minimize the confounding effect of all these variables, physiologists have defined 'Basal Metabolism', which is the rate of energy utilization by animals at complete rest, their tissues unstimulated by the digestion and assimilation of food or by low or very high temperature. Basal metabolism (BM) is commonly measured by the rate at which oxygen is consumed after a short period of fasting, so that food assimilation is not taking place. It is important to understand the concept of BM and how it is affected by other factors, since this is the foundation on which energy requirements for all other activities are superimposed. Basal metabolism shows a diurnal rhythm, being higher during the day in diurnal birds but higher at night in nocturnally active birds. Although many seabirds are active both by day and night, Adams and Brown (1984) found marked diurnal patterns of Basal Metabolic Rate (BMR) in the surface nesting Sooty Albatross and the burrow nesting Kerguelen Petrel. They chose to select the periods of lowest metabolic rate from these diurnal patterns as the best measure of BMR in these species (Figure 5.1).

The metabolic rate of a bird at complete rest and unstimulated by digestion is related to ambient temperature in a somewhat complex way. Over a range of warm temperatures the metabolic rate is

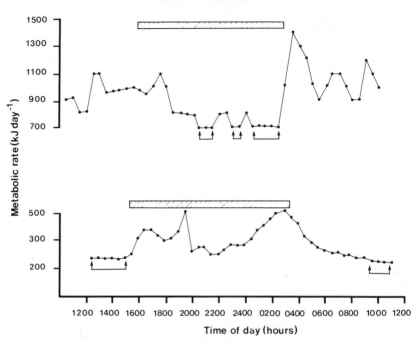

Figure 5.1 Metabolic rate of a surface-nesting procellariiform (Sooty Albatross) and a burrow-nesting procellariiform (Kerguelen Petrel) over 24 hours. Hatched areas indicate hours of darkness on the dates of the experiments. Stable periods of BMR are indicated with arrows. (From Adams and Brown, 1984.)

constant (and defined as the BMR). At higher temperatures, metabolic rate increases because body temperature rises as the bird is unable to lose heat quickly enough. At lower temperatures, metabolic rate increases with falling temperature since greater heat loss requires an increase in metabolic rate to maintain body temperature (Figure 5.2). The upper and lower critical temperatures (Figure 5.2) and the range of temperatures over which energy expenditure is constant at BMR (the 'zone of thermoneutrality') vary in relation to the size of bird. For smaller birds the lower critical temperature (the lowest temperature at which BMR can be sustained) is higher than it is for larger birds, since the rate of heat loss per gram of tissue decreases with increasing body size.

Although BMR can be strictly defined, some published measurements have been obtained using insufficiently stringent experimental procedures. Failure to consider circadian variation or to maintain

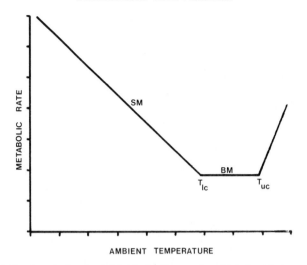

Figure 5.2 Basal and Standard Metabolic Rates of a bird in relation to ambient temperature. T_{lc} = lower critical temperature, T_{uc} = upper critical temperature of the zone of thermoneutrality. BM = Basal Metabolic Rate, SM = Standard Metabolic Rate. Scale values on the abcissa depend on the body weight of the bird. (From Kendeigh *et al.*, 1977.)

temperatures within the thermoneutral zone have led to overestimates of the BMR of some species (Adams and Brown, 1984).

The BMR of a seabird is related to its weight according to the allometric equation

$$M = aW^b \pm \text{s.e.e.}$$

This relationship is linear when expressed logarithmically

$$\log M = \log a + b \log W \pm \text{s.e.e.}$$

where M is the metabolic rate (expressed in calories or joules), W is the body weight of the bird, a and b are constants, and s.e.e. is the standard error of the estimate of M for a given value of W.

A number of equations of this form have been published over the last 25 years, with smaller standard error of the estimate as sample sizes increase and experimental procedures are improved. Kendeigh *et al.* (1977) showed that non-passerines follow a different relationship from passerines but that there are no major differences in the equations obtained for particular families of non-passerines. Thus the equation they give for non-passerines (based on 77 determinations) is

$$M = 2.1857W^{0.7347} \pm 5.176 \qquad (5.1)$$

(where M is in kJ and W in grams) and this can be used to predict the BMR of any species of seabird of known weight.

Warham (1971) and Grant and Whittow (1983) have suggested that procellariiform seabirds have lower metabolic rates in relation to body mass than do other seabirds, partly because the Procellariiformes appear to have unusually low body temperatures. The latter implies a lower metabolic rate unless the plumage of procellariiforms has an unusually high thermal conductance, which seems unlikely since most procellariiforms have dense, waterproof plumage. Croxall (1982) suggested that, within the procellariiforms, albatrosses and large surface nesting petrels may have a different relationship between weight and BMR from burrow-nesting petrels. If such differences did exist then it would be very much more difficult to estimate the food consumption by a seabird community composed of species from a variety of families for each of which we might require a different set of bioenergetics equations. Fortunately the available evidence suggests that the relationship between BMR and weight of procellariiforms is indistinguishable from the same relationship for other non-passerines (Figure 5.3). The slight discrepancy for three

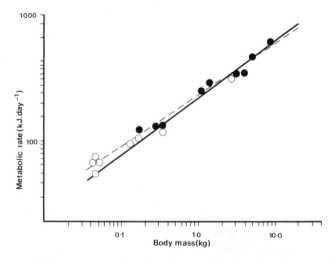

Figure 5.3 Relationship between metabolic rate and body mass of procellariiforms. ● sub-Antarctic species, ○ temperate and tropical species, solid line is predicted BMR from Kendeigh *et al.* (1977), dashed line is regression for actual data. (From Adams and Brown, 1984.)

of the smallest procellariiforms (Figure 5.3) is attributed by Adams and Brown (1984) to poor experimental technique, overestimating the BMR of these species, rather than species-specific variations in the overall relationship between BMR and size.

Another potential problem arises from the possibility that the basal metabolic rates of tropical seabirds may be lower than those of temperate or high-latitude seabirds (Pettit et al., 1985). This is a reasonable idea, since some tropical terrestrial birds appear to have unusually low metabolic rates, and this has been explained as an adaptation to living in an environment where heat stress may be a problem. A lower Basal Metabolic Rate should mean that birds are less likely to suffer heat stress during activity. Pettit et al. (1985) determined BMR for eight species of tropical seabirds and compared their results with energy expenditures predicted by the equation for BMR given by Lasiewski and Dawson (1967), which was derived from studies of a wide variety of nonpasserine birds, from hummingbirds to ostrich. Two species had higher BMRs than predicted, while six were lower. Four species of procellariiform were included in this study, and all four of these had BMRs lower than predicted. These species also had lower daytime body temperatures (Pettit et al., 1985), but this could partly arise from the fact that these species are nocturnally active when breeding, whereas terns and tropicbirds studied are diurnal.

This study suggests that tropical seabirds probably do not, as a rule, have unusually low BMRs, but it again raises the possibility that procellariiforms may. However, the discrepancies between measured and predicted rates, while interesting, were generally in the region of $\pm$ 10–15%, so may not be too important as far as ecological considerations of total energy budgets of seabird communities are concerned. Ellis (1984) showed that the BMR of charadriiform seabirds (mainly gulls and auks) did vary with latitude, and so ambient temperature. High latitude charadriiform seabirds had BMRs that were up to twice as high as those of tropical charadriiform seabirds of the same body mass. These results contrast with those of Adams and Brown (1984), and it is not clear whether this apparent relationship between BMR and latitude is universal or not. Since Ellis (1984) also showed that many seabirds have BMRs higher than predicted by equations for non-passerines in general, and there are significant differences in the body core temperatures of different Orders of seabirds (Furness and Burger 1987), there may also be slight differences between BMRs of different Orders of seabirds. The Procellariiformes and Sphenisciformes appear to have BMRs equivalent to those of other non-passerines, while BMRs of the Charadriiformes and Pelecaniformes appear to be higher than expected.

Despite these uncertainties in the details of relationships between BMR and latitude or Order of seabirds, we can reasonably assume that equation (5.1) can be applied with an acceptable degree of accuracy to all seabirds, since most discrepancies between regions or species are relatively small. In the simplest bioenergetics models of food consumption by bird populations (e.g. Berruti, 1983; Summers, 1977) the authors simply multiplied the BMR by a 'correction factor' of between 2 and 5 to allow for the extra costs of free living over BM. Since the multiplier is an arbitrarily chosen number, this procedure is unsatisfactory and does not inspire confidence in the results! It is preferable to determine a set of bioenergetics equations for each of the activities that form part of a seabird's daily life. The major ones are existence, gliding and flapping flight, swimming (including diving), egg-laying, incubation, moult and weight change.

5.2.4 *Existence Metabolism (EM)*

Existence Metabolism (EM) is the sum of Basal Metabolism, costs of thermoregulation, digestion and limited locomotor activity (i.e. excluding flight, swimming or running). EM has been estimated for a large number of bird species by measuring over a period of days the amount of energy consumed as food and subtracting the amount excreted unassimilated (from the intestine and kidneys) by caged birds maintaining constant weight and in controlled conditions of temperature and photoperiod. Compared with BMR, EM sacrifices physiological precision for ecological utility. EM provides a more convenient basis for bioenergetics modelling.

EM is greater for birds kept in longer photoperiods, and decreases linearly with increasing temperature up to the upper critical temperature of the zone of thermoneutrality of the BM. As with BM, there is little variation between different non-passerine families, and it is adequate to use one set of equations for all seabirds. However, the validity of EM equations as a basis for bioenergetics models has been questioned. Although allowing for differences in temperature, EM equations ignore other factors affecting heat exchange such as insolation, wind, black-body radiation or convection. Weathers *et al.* (1984) have suggested that models based on heat transfer theory might provide more accurate estimates of energy demands than those generally in use which are based on EM equations. Weathers (1979) also suggests that the metabolic rate of tropical seabirds may be lower than that of seabirds at higher latitudes, as an adaptation to reduce heat loads and the amount of water that must be used in evaporative cooling (Wiens, 1984). Adams and Brown's (1984) study of the BMR of procellariiforms does not support this idea, so the equations of Kendeigh *et al.* (1977) are probably reasonably accurate for all seabirds:

$$30°C, \text{ winter } n = 40, \ M = 6.088W^{0.6256} \pm 4.966 \quad (5.2)$$

$$30°C, \text{ summer } n = 70, \ M = 4.469W^{0.6637} \pm 5.523 \quad (5.3)$$

$$0°C, \text{ winter } n = 40, \ M = 17.719W^{0.5316} \pm 4.929 \quad (5.4)$$

$$0°C, \text{ summer } n = 70, \ M = 17.330W^{0.5444} \pm 4.715 \quad (5.5)$$

(weights W in grams, M (existence metabolism) in kJ).

Linear interpolation between these temperatures, or extrapolation below 0°C, gives the EM for a seabird of any given weight.

5.2.5 *Flight and swimming costs*

For seabirds, EM is generally the largest part of the daily energy budget, and costs of foraging activity are next in importance. Unfortunately, it is extremely difficult to obtain direct measurements of metabolic costs of flight or swimming by seabirds. Relationships for flight derived from aerodynamic theory (Pennycuick, 1969; Tucker, 1973) do not appear to fit with the few empirically determined results (Berger and Hart, 1974; Kendeigh *et al.*, 1977). King (1974) estimated flight costs of non-passerines to be 15.2 times BM, regardless of body weight. However, gliding flight is very much less expensive than flapping flight. Baudinette and Schmidt-Nielsen (1974) estimated the cost of gliding flight in Herring Gulls in a wind tunnel using a mask over the beak to obtain measurements of oxygen consumption, and obtained a value of 1.85 times EM. Flint and Nagy (in Wiens, 1984) used $D_2^{18}O$ techniques (see section 5.4) to determine that the metabolic rate of free-living Sooty Terns during flight (predominantly flapping flight) was 4.8 times BM. This is only one-third of the cost predicted by King (1974) which Flint and Nagy attributed to the high aspect ratio and low wing loading of Sooty Terns, suggesting that species-specific aerodynamic adaptations may play an important role, so making general models of flight costs unreliable.

Albatrosses possess a special locking mechanism in the wing to allow them to glide continuously, without muscular effort to hold the wings in position (Pennicuick, 1982), and this will clearly reduce their flight costs. Frigate-birds possess a unique layer in the pectoral muscle which is thought to allow them to soar at minimal muscular cost (Kuroda, 1961). It is reasonable to assume that seabirds with wings partly adapted for flight underwater (e.g. auks and diving petrels) will pay a penalty in terms of higher costs of flight in air because of their high wing loadings. This is an area where more research is required. At present we have to make the best we can of the available data. Croxall et al. (1984) used what they acknowledged to be a conservative value of 1.85 times EM in estimating costs of flight in sub-antarctic seabirds. Furness and Cooper (1982) estimated from various published sources that flapping flight of seabirds added $1.4 W^{0.67}$ kJ per hour to existence requirements, and gliding flight added $0.09 W^{0.73}$ kJ per hour to existence costs.

Less is known of the costs of underwater and surface swimming. Furness and Cooper (1982) assumed, for lack of data, that these activities cost the same as flapping flight and gliding flight respectively. Croxall et al. (1984) used empirical data from Kooyman et al. (1982) and Davis et al. (1983) to obtain an estimate for penguin foraging costs of 2.45 times EM. Since many seabirds feed by pursuing prey underwater, further studies of the energetic costs of diving will make an important contribution to more accurate estimation of seabird energy budgets.

Nagy et al. (1984) have used $D_2^{18}O$ to measure field metabolic rates of breeding Jackass Penguins, for which time-activity budgets were known, and some of which carried speed-time meters allowing their foraging distances to be measured. They estimated that brooding a chick cost 1.7SMR, foraging cost 6.6SMR, while underwater swimming cost 9.8SMR. These figures give results for overall daily energy budgets that agree well with estimates obtained earlier using bioenergetics equations. Daily energy budget computed by Nagy et al. (1984) from their $D_2^{18}O$ data was only 3% greater than an estimate derived from bioenergetics modelling by Furness and Cooper (1982), giving some support for the general usefulness of indirect modelling approaches.

5.2.6 *Egg production*

Egg production by seabirds may be constrained by nutrient demands rather than energy, as indicated for geese (Ankney and MacInnes, 1978) but in most seabirds, clutch sizes are small, and only in storm petrels does the weight of the single egg represent a large proportion of body weight. Gulls of several species can lay many more eggs than normal if these are sequentially removed before clutch completion, suggesting that nutrients do not limit egg production. Costs can be modelled as number of eggs × mean egg weight × calorific value × efficiency of conversion of body tissues into eggs.

Modelling of this sort suggests that egg production costs are rather a small part of the energy needs of a seabird population. Furness and Cooper (1982) found that the modelled egg production costs of Jackass Penguins, Cape Gannets and Cape Cormorants represented less than 0.1% of the total annual energy demand of the population. Croxall and Prince (1982) simply discounted egg costs as trivial in their calculation of the food consumption of South Georgia seabirds.

5.2.7 *Incubation costs and chick energetics*

In many seabirds, embryonic development is prolonged, incubation temperatures are lower than in other non-passerines and incubation occurs in long bouts during which the incubating bird is inactive and fasting. For these reasons it is likely that the energy cost of incubation is low. Croxall and Ricketts (1983) used the rate of weight loss of incubating albatrosses to estimate that incubation cost about 1.2 times BM. For penguins, weight loss during incubation bouts suggest a cost of 1.3 BM (Croxall, 1982). Using measurements of oxygen consumption, Brown and Adams (1984) estimated that incubation cost 1.4 times BM in Wandering Albatrosses and 1.0 times BM in penguins. Incubation costs of seabirds are clearly small, and for many species that have long bouts they may actually be less than existence metabolism because incubating birds are inactive and fasting.

Kendeigh *et al.* (1977) provide an equation for the energy budget of chicks in relation to their weight, which Furness and Cooper (1982) used in their seabird model. Croxall and Prince (1982) preferred to use direct field measurements of food intake on the grounds that empirical data are better than a generalized equation based on a small number of species from disparate families. This is discussed further by Wiens (1984).

5.2.8 *Moult costs*

Croxall (1982) calculated energetic costs of moult in penguins from weight losses during moulting fasts and estimated that moulting cost 2 times BM. In other seabirds moult occurs over a much longer period during which the birds continue with normal activities rather than fasting, so the costs will be spread more thinly than in penguins. No data exist for the costs of moult in seabirds. Feather replacement in the House Sparrow costs 774 kJ (Kendeigh *et al.*, 1977), and according to Turcek (1966) plumage mass is proportional to body mass to the power 0.96, and represents about 6% of body mass. Kendeigh *et al.* (1977) argued that if these data apply to all birds then we can generalize to the equation

$$\text{Moult cost (kJ)} = 34.7 \text{ weight (g)}^{0.96}.$$

However, Turcek does not state clearly what he defined as plumage. Other studies have found that flight feathers represent about 6% of body weight and that all other feathers (contour feathers important for thermoregulation) represent another 6% (Furness and Burger, 1987). Turcek was probably considering only the latter feathers, so that Kendeigh *et al.*'s equation probably applies to only 50% of the total feather mass. Masses of all feathers plucked from a variety of species are shown in relation to total body weight in Figure 5.4. Probably we need to double the value calculated in previous studies where the above equation has been used. However, this assumes that seabirds moult all their feathers at each moult, which is certainly not true for many species. For a variety of small passerines, moult costs about 450 kJ per gram of feather produced (Walsberg, 1983) so we may make a rough estimate of the energetic cost of moult in seabirds if we assume that this figure applies. However, the total weight of the plumage of seabirds is not well known, and not all feathers are replaced annually; some seabirds have rather complicated moult patterns which makes this calculation less accurate than it might appear. Fortunately, moult costs appear to be small in relation to costs of existence and foraging activity, so models that ignore moult costs altogether will underestimate food demands by only a very slight amount.

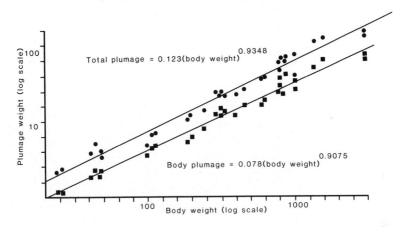

Figure 5.4 Flight feather and contour feather masses of a variety of seabirds in relation to total body mass.

5.2.9 *Assimilation efficiency*

Not all the energy in food is assimilated. Assimilation efficiency varies according to food type and bird species, but is generally between 79 and 82% for piscivorous birds (Ricklefs, 1974). Thus all calculations of energy requirements derived from studies of oxygen consumption need to be multiplied by 100/80, or 1.25, in order to give the amount of energy that must be ingested.

5.3 Simulation modelling

Having considered the various ways in which seabirds use energy, we are now in a position to consider how the overall budget can be estimated. Models of food consumption by seabirds combine estimates of population size, activity budgets and ambient temperatures with the bioenergetics equations described above. Numerous input parameters need to be provided and many of these have to be crudely estimated at our present level of knowledge. Models have undoubtedly improved rapidly since this approach began, but they can still often be criticized for providing results of unspecified accuracy. The obvious aim of these models is to provide as accurate an assessment as possible of food consumption, but one immediate result is that they highlight areas where precise data are lacking.

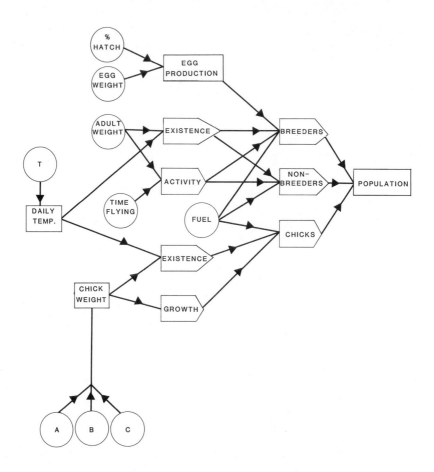

Figure 5.5 Population bioenergetics model showing input parameters and functions for which these are required. (From Furness, 1982.)

This is an important and valuable result of modelling which should lead to further research being done in the areas where data are needed.

The structure of a typical bioenergetics model for seabirds is shown in Figure 5.5.

In order that we can use models with confidence, we need to know how sensitive their predictions are to variations in each of the input values, and also how close to reality their predictions are. These two topics, model sensitivity and model validation, are areas in which research activity is still at an early stage.

5.3.1 Model sensitivity

The sensitivity of a model may be defined as the percentage change in the output estimate as a result of a 1% change in the magnitude of one input value. Determination of sensitivity values should be carried out whenever this kind of simulation modelling is undertaken, as it indicates which parameters must be known precisely to give output results with small errors.

In a model used to estimate the fish consumption by seabirds of Oregon (Wiens and Scott, 1975) this type of sensitivity analysis showed that small alterations of input parameter values generally had a correspondingly smaller, often negligible, influence on the model output estimates of total breeding season energy demands (Wiens and Innis, 1974). However, the parameters with the highest sensitivity values depend on the characteristics of the species as well as on the structure of the model. For example, a model applied to a Great Skua population was most sensitive to the estimation of existence metabolism, population size and assimilation efficiency, while the same model applied to an Arctic Tern population was most sensitive to the activity budgets of adults and population size (Furness, 1978). This comes about because the Arctic Tern spends a high proportion of the day foraging, whereas the Great Skua spends most of the day inactive on its territory. In contrast, since the costs of egg production are small in relation to the costs of existence and activity, the sensitivity of the model to the input value of egg weight is about 500 times less than sensitivity to parameters such as population size or assimilation efficiency.

What happens if, as in the real world, all the input values are in error to a greater or lesser degree? In some cases errors will have additive effects on the output values, in others they will tend to cancel out. The extent to which this occurs is important, and can by investigated by using a Monte Carlo technique, so called because it relies on the use of numerous runs using random number distributors to modify the input values. Furness (1978) used a computer function to generate a random value for each parameter, with a specified normal distribution using the known mean and standard error for each parameter. The errors in estimated parameter values were assumed to be uncorrelated and the generated set of parameter values was then input into the model. Repeating this process 300 times gave 300 estimates of the annual food requirements of a population, and so allowed the standard deviation and confidence limits to be calculated. Applying this to models constructed in the 1970s shows that they have rather a wide 95% confidence interval of ± 50% of the mean, while models constructed in the 1980s using the better bioenergetics equations have a confidence interval of about ± 30%.

5.3.2 *Model validation*

It is a weakness of the modelling approach that, as yet, no detailed validation has been possible. Validation requires simultaneous and independent direct assessment of metabolic costs using techniques such as doubly-labelled water or physiological telemetry (described in section 5.4). Such studies have just begun, but offer encouragement that bioenergetics models give reasonable results (Wiens, 1984; Bryant and Westerterp, 1983). Comparison of doubly-labelled water and bioenergetics estimates for passerine birds suggest that models may overestimate requirements by 5% to 20% (Wiens, 1984), but no such comparisons are yet available for seabirds. Until they are we must be cautious in the interpretation of results from bioenergetics models. Some support for the use of bioenergetics modelling has been provided by a study of Jackass Penguin energetics carried out using doubly-labelled water (Nagy *et al.*, 1984) where results from this method were within 3% of estimates obtained earlier for the same population by Furness and Cooper (1982), but this agreement could be fortuitous and needs to be checked by simultaneous labelled water and bioenergetics estimates.

5.3.3 *Applications of modelling*

Models of seabird energetics are mostly used to estimate energy flow into seabird populations occupying a defined area over a specified period of time. This can be either the foraging range from a particular breeding colony over part or the whole of a breeding season, or a restricted area of sea or ocean at any time of year. In the first case estimates of numbers of pairs breeding at the colony and associated nonbreeders form the basis for the calculations. In

Table 5.1 A list of bioenergetics models used to calculate food consumption by populations of seabirds

Evans (1973)	North Sea	Metabolic multiple	Area
Wiens and Scott (1975)	Oregon coast	Integrative	Area
Everson (1977)	Southern Ocean	Metabolic multiple	Area
Furness (1978)	Foula, Shetland	Integrative	Colony
Grenfell and Lawton (1979)	Southern Ocean	Metabolic multiple	Area
Mougin and Prevost (1980)	Southern Ocean	Metabolic multiple	Area
Croxall and Prince (1982)	South Georgia	Integrative	Colony
Furness and Cooper (1982)	Benguela region	Integrative	Area
Schneider and Hunt (1982)	Bering Sea	Metabolic multiple	Area
Ford et al. (1982)	Pribilof Islands	Integrative	Colony
Croxall et al. (1984)	South Georgia	Integrative	Colony
Wiens (1984)	Kodiak Island	Integrative	Colony
Wiens (1984)	Pribilof Islands	Integrative	Colony
Guillet and Furness (1985)	Dassen Island	Integrative	Colony
Montevecchi (unpubl.)	Newfoundland	Metabolic multiple	Colony
Furness and Barrett (1985)	N. Norway	Integrative	Colony
Gaston and Diamond (unpubl.)	Eastern Canada	Integrative	Area
Hunt (in press)	SE Bering Sea	Metabolic multiple	Area
Hunt (in press)	S. Orkney Islands	Metabolic multiple	Area
Schneider et al.(in press)	SE Bering Sea	Metabolic multiple	Area

the second, estimates of the densities of seabirds at sea provide the basis. This distinction is important since the colony-based approach, by focusing on an important seabird colony, tends to overestimate the global impact of seabirds on their prey populations. The question is whether the aim is to investigate the trophic ecology of a seabird colony, or the impact of seabirds on prey populations. In the latter case area-based estimates are preferable to ones from studies of individual colonies. Bioenergetics models applied to seabird populations are listed in Table 5.1.

Integrative models (see 5.2.2) can be used to examine the seasonal pattern of energy demand (Figure 5.6). As shown in Figure 5.6, the energy demand can be illustrated separately for different classes within the population. This can provide some insights into seabird breeding biology. For example, the peak amount of food required for chicks never equals the needs of the breeding adults in Figure 5.6. Alternatively, the patterns of allocation of energy can be compared. Over a year, estimated costs of egg production and moult of three species of South African seabird represented only about 0.1% and 2% of the total budget, while 50% to 71% went to adult existence and 19% to 36% to adult foraging costs (Table 5.2). True figures may differ a little from these estimates, since several parameters in this model were only roughly estimated, but these results indicate an

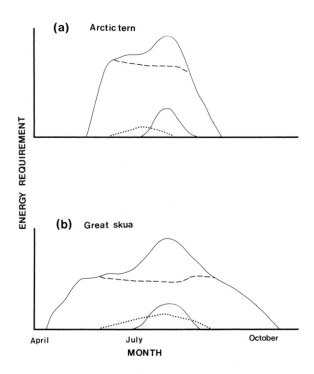

Figure 5.6 Seasonal pattern of population energy requirement for Arctic Terns and Great Skuas as a Shetland colony as estimated by bioenergetics modelling. (From Furness, 1978.)

Table 5.2 Annual energy requirements for adult existence, foraging locomotion costs, chick existence plus growth, moult and egg production for seabird populations in the Saldanha region of the Benguela upwelling (from Furness and Cooper, 1982).

Category	Jackass Penguin	Cape Gannet	Cape Cormorant
Total energy requirement (kJ × 10^8)	321.7	201.2	113.4
Percentage for: Adult existence	71.4	50.7	63.6
Adult foraging	19.4	36.1	28.5
Chicks	7.6	11.0	5.4
Moult	1.5	2.1	2.5
Egg production	0.1	0.1	0.1

overwhelming dominance of existence costs, and the minor nature of costs of moult and egg production.

Furness (1978) estimated the total energy demand of the seabirds at Foula, Shetland to be 5×10^{10} kJ per year. Wiens (1984) estimated that the 11 species of seabirds at the Pribilof Islands had a total breeding-season energy demand of 2.7×10^{11} kJ. Croxall and Prince (1982) made a preliminary estimate that seabirds at South Georgia consumed nearly 10^{13} kJ during their breeding season, or two hundred times the amount taken by seabirds at Foula. However, these figures can only be put into context when the diets and foraging ranges of the seabirds are also considered, along with the production of the prey organisms.

Wiens and Scott (1975) estimated that guillemots off Oregon consumed 5700 tonnes of herring, 3500 tonnes of anchovies and 3000 tonnes of smelt annually, and that all the seabirds consumed about 22% of the net annual production of small pelagic fish within 185 km of the coast. Furness (1978) suggested that consumption by seabirds at Foula of 5×10^{10} kJ was equivalent to approximately 29% of the annual fish production within 45 km of the colony, but pointed out the wide confidence interval associated with this estimate. This figure has been challenged on the grounds that the main consumer species, the Fulmar, may travel more than 45 km to feed, though there is little evidence to support this claim, and on the grounds that fish migrations and oceanographic currents may lead to larger quantities of fish being available to seabirds in that area. For such reasons, area-based studies are likely to give better estimates of seabird impact on prey. Evans (1973) made a pioneering estimate of the energy requirements of seabirds in the entire North Sea. He obtained an estimate that a minimum of 13% of annual fish production was taken by seabirds over the area as a whole. Furness and Cooper (1982) calculated that seabirds in the Saldanha fishery area of the Benguela current consume about 20% of anchovy and round herring biomass annually, but smaller amounts of the stocks of other species. Their predation on anchovy would therefore be equivalent to about one-third of the average quantity caught by the commercial fishery in this area.

In the south-east Bering Sea consumption by seabirds is largely of juvenile pollock. Consumption was estimated at 1.5×10^5 tonnes per year (Schneider and Hunt, 1982), which compares with commercial catches of 1.7×10^5 tonnes in 1964 and 18.7×10^5 tonnes in 1972 (Smith, 1981). Croxall *et al*. (1984) estimated that South Georgia seabirds consumed 6 million tonnes of krill, one million tonnes of copepods, nearly half a million tonnes of squid and fish and a quarter of a million tonnes of amphipods annually. Macaroni Penguins are responsible for more than half of the krill consumption and have a maximum foraging range of about 100 km. From this they can be computed to consume 123 grams of krill mer square metre per year within their foraging range, or 2.5 grams per cubic metre per year if their diving depth is taken into account. Unfortunately, these data cannot presently be related to standing stocks and production of food. Preliminary acoustic surveys for krill around South Georgia produced estimates of only about 0.5 million tonnes (Croxall *et al.*, 1984) which, in view of the calculated consumption by seabirds, must be far too low a figure. Very little is known of the stocks of squid, but it seems probable that seabirds are of major importance as consumers of krill and squid in the Southern Ocean.

Wiens (1984) combined bioenergetics estimates of energy demand with data on foraging distributions of seabirds around the Pribilof Islands in order to demonstrate the importance of each species at varying distances from the colonies (Figure 5.7). Guillemots accounted for most food consumption within 70km while Fulmars were dominant beyond 120 km. Altogether, most of the energy demand occurred within 50 km radius of the islands (Figure 5.7).

Ford *et al*. (1982) adapted bioenergetics modelling to assess the potential impacts of changes in food resource levels on seabirds, by combining bioenergetics considerations with central-place foraging theory. They assumed that breeding seabirds should minimize the time spent in flying to and from foraging grounds to maximize the rate of food delivery to the young, then looked at the influences of disruptions in the abundance or availability of food in such situations. This method provides a useful tool for investigating interactions between energy demands, foraging behaviour and aspects of breeding ecology such as chick growth rates, age at fledging, breeding success and adult activity budgets (Wiens, 1984).

From the modelling described in this section we can conclude that food consumption by seabirds can be considerable, at least in some ecosystems, and so interactions with other components of the system, such as fisheries, seem likely to occur.

5.4 Direct measurement of seabird metabolism

Heart-rate biotelemetry may be of use in measuring free-living metabolism. It has been used to study the diving and flying respiration of unrestrained birds (Butler, 1980) and to estimate metabolic costs of Redshanks in laboratory conditions (Ferns et al., 1980). Its application to field studies is hindered by individual variations in the relationship between heart rate and oxygen consumption (Ferns et al., 1980), changes in heart stroke volume or oxygen content of blood independent of heart rate (Butler et al., 1977), changes in heart rate induced by emotional stresses during periods when the metabolic rate may remain unchanged (Ball and Amlaner, 1980), and by UK Home Office regulations regarding the liberation into the field of birds carrying heart-rate transmitters.

The most useful method is likely to be the use of doubly-labelled water ($D_2^{18}O$) to determine the total energy budget over a period of time between injection and recapture for removal of a body-water sample (Lifson and McClintock 1966). This requires the capture and recapture of an individual over a period of one or two days, which is not always easy to achieve. It assumes that the bird's behaviour is not altered by capture and injection, and involves the use of costly isotopes and technically complicated laboratory analyses of samples. Nevertheless, the method has been applied with great success to examine the energetic costs of breeding and foraging in birds (see for example Bryant, 1979; Bryant and Westerterp, 1983). One problem with applying this method to seabirds is that the quantity of $D_2^{18}O$ which needs to be injected is proportional to body mass. Seabirds tend to be large compared to the passerines on which the technique has mostly been used, making its application to seabirds very expensive. Doubly-labelled water provides a powerful technique to validate bioenergetics models for seabirds and to examine the energy expenditures of individual seabirds in particular circumstances (for example in relation to brood size or chick age).

5.5 Changes in ecosystem structure

May et al. (1979) warned of the problems that face those who wish to understand the effects of the exploitation of marine ecosystems: 'multispecies ecosystems will often manifest complex "catastrophic" behaviour. This transformation will not usually be continuously

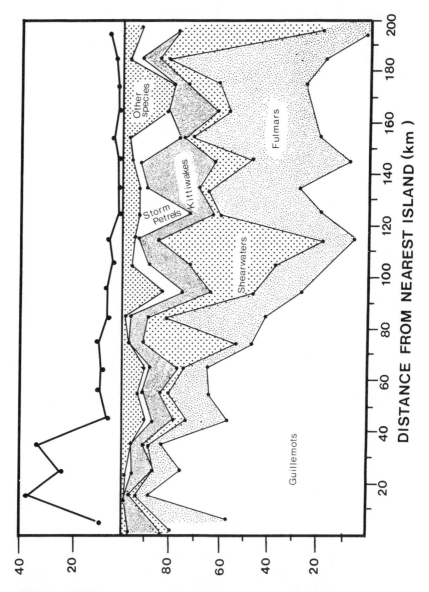

Table 5.3 Breeding biology of Peruvian guano seabirds compared to related seabirds from other areas. Data from Nelson (1978) and Cramp *et al*. (1974).

Taxon	Character	Peru	Elsewhere
Boobies		Peruvian Booby	All other Boobies
	Mean clutch size	2.5	1 to 2
	Chick weight gain per day (g)	40	15 to 40
	Brood weight gain per day (g)	100	15 to 80
	Fledging period (days)	98	100 to 170
	Age at first breeding (youngest)	2	3 to 5 or more
Cormorants		Guanay Cormorant	All other cormorants
	Mean clutch size	4	average 3

reversible. Such changes are seldom, if ever, predictable in a quantitative sense'. Thus, while we have seen that seabirds do form an important part of the upper trophic levels of marine ecosystems, we cannot make quantitative or even reliable qualitative predictions of the changes that will occur as a result of the perturbation of the system. To complicate the issue further, there is considerable dispute among marine and fisheries biologists over the relative roles of long-term oscillations in oceanography or climate and of fisheries in bringing about changes in the patterns of energy flow in marine ecosystems.

While many changes in fish stocks and ecosystem structure have been observed, it is fair to say that the causal factors involved are largely a matter of speculation and controversy. For this reason we shall consider first some examples in regions of upwelling of nutrient-rich cool water, since the food chains in such areas tend to be short and simple and so provide fewer complex links within the ecosystem.

5.5.1 *Upwelling ecosystems*

Coastal Peru is one of the main regions of upwelling in the world (Menzel *et al*., 1971). Cool, nutrient-rich waters of the Humboldt current rise to the surface to promote rapid growth of large phytoplankton which are fed upon directly by anchovies. The anchovy

Figure 5.7 Spatial distribution of total community energy demand (upper section) and the percentage contribution of each species or species group to that total, as functions of distance from the colony for seabirds in the Pribilof Islands, Bering Sea. (From Wiens, 1984.)

supported a fishery which began in the 1950s, and grew to become the largest single-species fishery in the world, producing 10 million tonnes during the year 1968–69. Anchovies also provide the staple diet of an enormous seabird community, consisting of millions of Peruvian Cormorants and boobies, and smaller numbers of Brown Pelicans, penguins, petrels, gulls and terns.

Seabird populations are severely affected by oceanographic perturbations which occur, for reasons that are not yet understood, about every seven years. Upwelling intensity varies seasonally, being weakest in the southern hemisphere's summer, when warm oceanic water (which is nutrient-poor) can displace the denser, cold nutrient-rich water. In most years this effect is small, but occasionally it is pronounced and is called El Niño. This can have catastrophic effects on the ecosystem. The rise in water temperature during an El Niño results in anchovy shoals dispersing. This causes breeding failures and high mortality among the seabirds on a scale unknown anywhere else in the world. These semi-regular catastrophes must have exerted tremendous selection pressure on the seabird populations, favouring the ability to increase rapidly in numbers after each crash. The Peruvian Booby and Peruvian Cormorant have larger clutches than related species, may attempt to breed more than once within a year, and reach sexual maturity at an unusually early age (Table 5.3). These characteristics are probably selected because food becomes superabundant in the period following each crash. At that time even young, inexperienced, adults are able to raise large broods because the food supply per bird is much greater than for a population which has reached equilibrium with the environment. In this respect, Peruvian seabirds are most unusual in being comparatively r-selected, since most seabirds show the characteristics of K-selected species (MacArthur and Wilson, 1976).

If we examine the changes in numbers of all Peruvian seabirds over this century we can see that the introduction of the huge anchovy fishery had a devastating effect on the total guano-bird numbers, and their importance in the overall ecosystem. Seabird numbers can be assessed from their annual guano production. On average, one bird deposits 15.9 kg of harvestable guano per year (Jordan, 1967). Between 1909 and 1962, numbers of guano birds estimated from guano yield fluctuated widely, recovering from each crash induced by El Niño events (see Figure 4.3). The generally increasing trend in numbers in the early part of this century can be explained as a

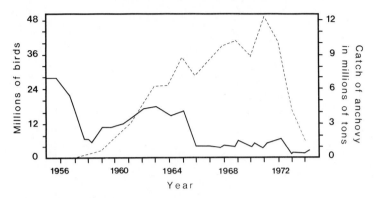

Figure 5.8 Numbers of Peruvian guano birds (counted) in relation to annual catches of anchovies. (From Santander, 1980.)

recovery of the birds from excessive guano-cropping disturbance and direct persecution of adult birds in an early period of uncontrolled exploitation; from 1909 the birds received protection of the Guano Administration and were provided with extra breeding areas (Duffy, 1983). Counts of guano birds between 1955 and 1974 were compared with annual anchovy catch data (Figure 5.8) by Jordan and Fuentes (1966), Schaefer (1970) and Santander (1980). This figure indicates that the guano birds did not recover quite as rapidly as normal after the 1957–58 crash, and this resulted in the numbers falling in 1965–66 to a record low since 1915, and then falling even further in 1972. The 1963 and 1965 crashes reduced the total population from 17 million to 3–4 million birds, a mortality of 76–82%. After these crashes numbers hardly recovered, remaining well below 5 million individuals in 1966–67 and 1967–68 (Schaefer, 1970). The next El Niño in 1972 reduced the population of guano birds to about 2.5 million and no increase in numbers was seen after this event. Thus, since the establishment of the anchovy fishery, the dynamics of the Peruvian guano seabird populations have changed. Instead of rapidly increasing by raising large broods at least once each year, they failed to respond to the reduced competition brought about by their reduction in numbers. The reason for this seems to be that the anchovy fishery has taken up the superabundance of food on which the Peruvian guano seabirds depended in order for them to cope with the recurring crashes induced by oceanographic perturbations.

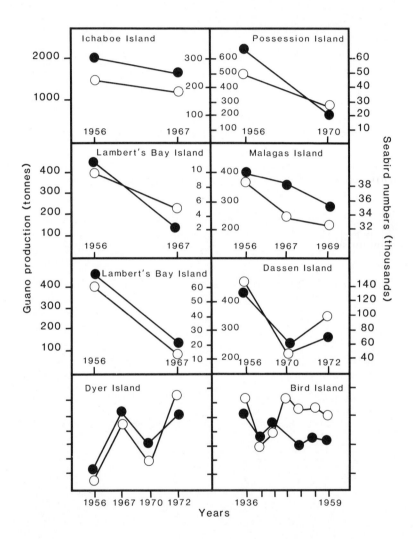

Figure 5.9 Relationships between seabird population size (○) and guano harvests (●) for a number of sites off Namibia and South Africa. (From Crawford and Shelton, 1978.)

Jordan and Fuentes (1966) estimated from field data that each guano seabird consumed on average 430 g of anchovy per day. Bioenergetics considerations would indicate that a Peruvian Cormorant, weighing about 2000 g, at about 20°C would require 200 g of anchovy for existence metabolism, and perhaps twice this amount to cover the additional costs of foraging, moult, egg production and chick rearing, so the estimate of 430 g per day is probably about right. From this value, Schaefer (1970) estimated that in 1961–65 the 17 million guano seabirds consumed 2.6 million tonnes of anchovy per year, or about one-quarter to one-third of the amount then taken by man, while after the 1965 crash the population of 4.5 million seabirds consumed 0.7 million tonnes per year from 1965 to 1968, which was less than 10% of the amount taken by man. Schaefer (1970) estimated that the maximum sustainable yield of anchovies in the early 1960s was about 7.5 million tonnes per year, but the catch taken by man was over 8 million tonnes each year between 1963 and 1968. After the crash in seabird numbers in 1965 the maximum sustainable yield estimates were higher, at around 8 million tonnes, and Schaefer (1970) attributed this change to the reduced consumption of anchovies by seabirds. Catches of anchovies by the fishery exceeded this new maximum sustainable yield in each year from 1966 to 1971, before crashing in 1972 and 1973, almost certainly in this case as a result of overfishing combined with the effects of the 1972–73 El Niño.

It is quite clear that Peruvian seabirds were unable to compete with the commercial fishery for anchovies, as shown by the progressive reduction in seabird numbers at each El Niño since the fishery began in the 1950s. Schaefer's (1970) analysis also suggests that the Peruvian seabird populations of the 1960s had a detectable influence on the potential yield of anchovies to man (as indicated by the calculated maximum sustainable yield).

An analogous upwelling to that off Peru occurs off South Africa and Namibia, in the Benguela current system. As in Peru, seabird guano has been harvested annually at colonies. Crawford and Shelton (1978) found a good correlation between guano production and known seabird population sizes (Figure 5.9), and inferred that the guano yield is largely determined by the number of breeding pairs of guano seabirds (predominantly Cape Cormorants, Cape Gannets and Jackass Penguins). They went on to compare seabird population changes measured from guano yields with estimates of fish stock abundance, obtained from catch, catch-per-unit-effort, or virtual population analyses. Adult pilchard provide the main food for Cape Gannets breeding on Ichaboe Island and form the bulk of catches landed at the nearby port of Luderitz. Guano production on Ichaboe Island correlates well with estimates of the biomass of the adult pilchard stock (Figure 5.10), suggesting a close dependence of seabird populations on their food resource. In Lambert's Bay the guano seabirds feed on juvenile pilchards which recruit into the commercial fishery in this area. Guano production and the estimated biomass of O-group pilchards in Lambert's Bay (Figure 5.11) are highly correlated, as in the previous example. Crawford and Shelton (1978) were also able to demonstrate that predatory fish show the same relationships to the abundance of small pelagic fish as do seabirds. The catch-per-unit-effort of snoek, one of the main fish predators of anchovies and pilchards, showed a high correlation with guano production by seabirds breeding at colonies nearby. Throughout the period before commercial fishing for pelagic fish, Crawford and Shelton found a pattern of guano production peaks at approximately 30-year intervals, and suggest that there may be

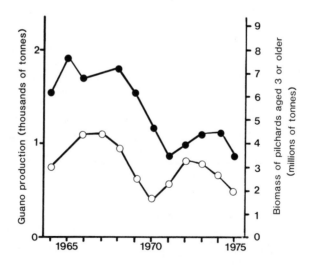

Figure 5.10 Relationship between guano production (●) on Ichaboe Island and biomass of pilchards aged three or older (○). (From Crawford and Shelton, 1978.)

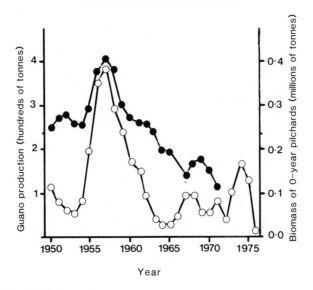

Figure 5.11 Relationship between guano production (●) on Bird Island, Lambert's Bay and biomass of O-group pilchards (○). (From Crawford and Shelton, 1978.)

a long-term cycle in the Benguela current affecting fish production. Since the development of the pelagic fishery in the Benguela current, the stocks of pilchards have been reduced and seabirds have become more dependent on anchovies. Seabird numbers have fallen as pelagic fish stocks have been reduced by intensive fishing, and the Jackass Penguin has been affected far more than the cormorants or gannets. The probable reason for this is the flightlessness of the penguin, making it more vulnerable to changes in the predictability and uniformity of fish distributions (Frost *et al.*, 1976).

5.5.2 *The Southern Ocean*

The Antarctic krill *Euphausia superba* is the dominant organism in the second trophic level of the Southern Ocean ecosystem. It supports a complex community of consumers: fish, small cephalopods, baleen whales, seals and seabirds. The fish and squid in turn support sperm whales, seals, larger cephalopods and other species of seabirds. Energy flow through this ecosystem is predominantly

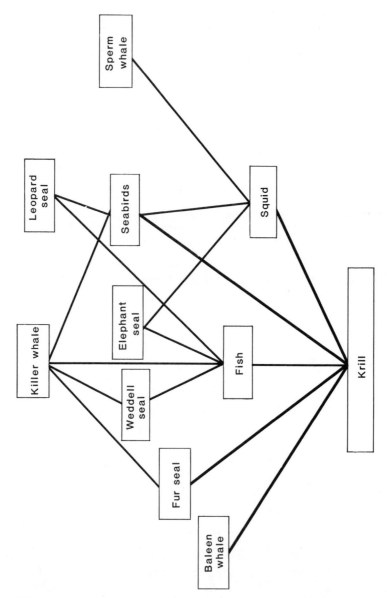

Figure 5.12 Simplified food-web for the Antarctic marine ecosystem.

determined by stocks of baleen whales, crabeater and fur seals, squid, and penguins (Figure 5.12).

In recent years there has been an upsurge of interest in this eco-system due largely to the realisation that krill is a highly productive resource which could support a very large commercial fishery. Krill production probably exceeds the recent total world fish catch by a factor of three or more (Everson and Ward, 1980) so there is a strong incentive to develop economic catching techniques. A further stimulus to initiate krill fishing comes from the suggestion that the depletion of baleen whale stocks caused by excessive exploitation will have resulted in a 'krill surplus' available for harvesting.

Laws (1977) noted that whales show segregation in Antarctic waters. Within species, the migration of different classes is staggered in re-lation to size and feeding requirements. Larger individuals tend to reach higher latitudes and pregnant females arrive before lactating ones. In addition, the larger and older whales arrive first and tend to occupy central, presumably optimal, parts of the feeding grounds. This suggests competition for food which might have been limiting to whale stocks. Since stock reduction there is evidence that food resources have improved for individual whales. Pregnancy rates of blue whales and fin whales have increased. Pregnancy rates of sei whales also increased, but this preceded the large-scale exploitation of that species, indicating that it was not a response to changes in the social structure or density-dependent behaviour of the population as a result of exploitation; rather it supports the view that it resulted from the effect of whaling on the food supply for baleen whales in general (Gambell, 1973). Growth rates of whales have increased, and age at sexual maturity has decreased, and these changes are also best explained in terms of an improved food supply per individual as a result of stock depletion. What effect has this stock reduction had on other species? It appears to have increased krill availabil-ity for seals and seabirds. Crabeater seals reached sexual maturity when four years old between 1945 and 1955, but when three years old between 1965 and 1970 (Laws, 1977). Many Antarctic seabird populations are increasing, apparently as a response to increased krill availability. King Penguins were heavily exploited in the nine-teenth century so their present increases may be partly a recovery from this. However, most Antarctic seabirds were not exploited, so their increases are not due to a history of early persecution and subsequent protection. More importantly, the seabird species showing the greatest rates of increase are those that feed most heavily on krill, while squid feeders are increasing rather slowly or not at all (Conroy, 1975; Croxall *et al.*, 1984).

The changes in the demographic parameters in seal and whale populations and the increases shown by populations of seabirds and seals, and particularly the greater response of the krill feeders, suggest that the populations of whales, seals and seabirds were all held at equilibrium sizes by competition for krill during the period before exploitation of the whale stocks. Reduction of whale stocks has improved krill availability and so upset the competitive balance between species. This can be seen in a model developed by May *et al.* (1979) which describes the response of interacting populations of seals, baleen whales and krill to three different imposed harvesting regimes. Although their model did not include seabirds, we can consider seabirds to respond in qualitatively the same way as seals since both these groups feed predominantly (in terms of energy flow) on krill but are not subject to harvesting regimes. Figure 5.13*a* shows the expected effect of stopping all whaling. Figure 5.13*b* shows the effect of maintaining whaling and also initiating a krill fishery. Figure 5.13*c* shows the effect of stopping whaling but initiating a krill fishery. All three scenarios would result in a reduction of seabird and seal populations, most drastically if whaling stopped and a krill fishery was started. Perhaps this last is the most likely of the three possibilities.

While these studies have indicated that krill availability is of major importance in the dynamics of the Southern Ocean seabird populations, we have not yet considered quantitative aspects of their

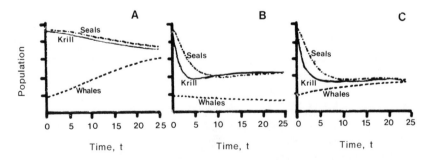

Figure 5.13 The behaviour of populations of krill (solid line), whales (dashed) and seals (dot-dash) in a simulation model under three different harvesting regimes: (*A*) After cessation of whaling, (*B*) with whaling and also exploitation of krill, (*C*) with exploitation of krill but no whaling. (From May *et al.*, 1979.)

role in the ecosystem and how this relates to their competitors, seals and baleen whales. Crude estimates of biomass show that for the Southern Ocean as a whole, the present biomass of whales still exceeds that of seals which exceeds that of seabirds. However, the energy requirements of seabirds greatly exceed those for the same biomass of seals or whales, since their metabolic rates are much higher, so consideration of biomass is misleading. Croxall *et al.* (1984) have used a bioenergetics model to estimate the food consumption of seabirds breeding at South Georgia, probably the biggest single aggregation of seabirds in the Antarctic. They estimated that the breeding seabirds of South Georgia consume annually some 7.8 million tonnes of food, of which 73% is krill, 13% copepods, 6% squid, 5% fish and 3% amphipods. Most of the food consumption is due to penguins, particularly Macaroni Penguins. From knowledge of the foraging ranges of seabirds from South Georgia they concluded that about 5 million tonnes of food was consumed in the South Georgia shelf waters alone. Croxall *et al.* (1984) were frustrated in their attempt to relate these estimates of consumption by seabirds to the stock biomass and production of the prey, since no reliable data were available. Indeed, the estimated consumption of krill by seabirds exceeds the current estimate for standing stock of krill in the South Georgia shelf waters by a factor of at least 6 and probably 10, while knowledge of squid stocks is largely due to sampling of squid beaks from predators such as albatrosses and whales.

The approach used by Croxall *et al.* (1984) cannot be applied to the entire Southern Ocean at present because knowledge of seabird population sizes, diets and activity budgets in other parts of the Southern Ocean is as yet inadequate. For example, Williams *et al.* (1979) estimated the world population of the Macaroni Penguin to be about 4 million breeding pairs, but Croxall and Prince (1979) gave a figure of 5.4 million pairs on South Georgia alone and stated that their census might underestimate by as much as 50%. Present sizes and age compositions of seal populations are rather better known, and fairly reliable data exist for whales. Assessing population sizes as they were in 1900, in order to estimate the change in krill consumption since whale stock reduction, is rather more difficult. Mackintosh (1973) estimated that baleen whale biomass had decreased by 36.5 million tonnes since 1900, from 43 to 6.5 million tonnes. Lockyer (1976) estimated that whales consume 3.5% of their body weight per day over a feeding season of about 120 days. This

D

indicates a consumption of 180 million tonnes of krill in 1900 and 28 million tonnes now, suggesting a 'krill surplus' resulting from whale stock depletion of about 152 million tonnes per year. Some of this is obviously now taken up by seals, cephalopods, fish and seabirds. Everson (1977) attempted to assess how much seabirds take by multiplying population size estimates by roughly estimated daily food intake data from field studies, an approach which we earlier found to be rather unreliable. He suggested that the total stock of seabirds in the Antarctic, which comprised 487 000 tonnes, mainly penguins, eat 15 to 20 million tonnes of krill, 6 to 8 million tonnes of squid and 6 to 8 million tonnes of fish each year. This equates to 1.4×10^{14} kJ per year. Using quite independent data, Mougin and Prevost (1980) combined estimates of Southern Ocean seabird populations with bioenergetics equations for existence metabolism multiplied by a factor of two to allow for activity. They obtained a total food consumption per year of 1.5×10^{14} kJ, or 39 million tonnes.

Grenfell and Lawton (1979) used a different bioenergetics approach to the same problem. They combined population census data with bioenergetics equations for resting metabolic rate and a correction factor of three (based on data presented by King (1974)) to allow for free-living costs. Although again a fairly crude calculation, they obtained an estimate of the annual consumption by seabirds of 1.4×10^{14} kJ, in close agreement with the other independent calculations. Grenfell and Lawton went on to consider this consumption by seabirds in terms of 'krill equivalents'. Thus, squid eaten by seabirds will have consumed a quantity of krill approximately five times as great as their own biomass. When this indirect consumption of krill by seabirds that feed on squid or fish is taken into account then annual consumption of krill equivalents amounts to between 100 and 140 million tonnes. This nearly equals Laws' (1977) estimate of krill consumption by baleen whales before exploitation (180 million tonnes) and greatly exceeds their current consumption (28 million tonnes).

Grenfell and Lawton (1979) made equivalent estimates of krill consumption (direct and indirect) by seals and whales in the Antarctic, using a bioenergetics equation for whales and seals

$$\text{Resting Metabolic Rate (kJ per day)} = 528.0W^{0.77}$$

They assumed that the actual metabolic rate in field conditions was about double this resting rate and also made small adjustments to

Table 5.4 Estimated krill consumption (direct plus indirect) by whales, seals and seabirds in the Southern Ocean in 1900 and the present (assuming Field Metabolic Rate = 2 × Resting Metabolic Rate and for whales production/assimilation = 30%) (from Grenfell and Lawton, 1979).

Predator	Present consumption of krill (10^6 tonnes)	1900 consumption of krill (10^6 tonnes)	Difference
Baleen whales	44	304	−260
Sperm whales	16	33	−17
Seals	185	80+	+100?
Seabirds	120	30+	+90?
Total (excluding fish and squid)	365	447+	−80?

allow for growth and for seasonal fat deposition and an assimilation efficiency of about 80%. Their resulting estimates of krill consumption (direct plus indirect) are given in Table 5.4. Consumption figures for the year 1900 for seabirds and seals can only be guessed at, since seal and seabird population sizes at that time are largely unknown. It seems likely, according to most authorities on Antarctic vertebrates, that seal numbers have more than doubled since 1900, so that their krill consumption in 1900 was probably no more than half the present amount. This means less than 100 million tonnes, and 80 million tonnes seems a reasonable guess. Some Antarctic seabird populations are increasing rather rapidly: Adelie Penguins at about 3% per year and Chinstrap Penguins at 8% per year at parts of Signy Island, and Macaroni Penguins at up to 9% per year at South Georgia (Croxall and Prince, 1979). If even much lower rates of increase apply to other colonies, Antarctic seabird numbers could have been doubling every few decades this century. In that case, their consumption of krill in 1900 may have been very much less than the present level. As a crude approximation we might suggest a figure of about one-quarter of the present level. Using these figures, which are rather approximate, we can see from Table 5.4 that the total consumption of krill at present by whales, seals and seabirds seems to be rather less than in 1900. In other words, seabirds and seals appear not to have taken up all of the 'krill surplus' left by the reduction of whale stocks. However, they have accounted for a substantial part of the surplus. The remainder has probably been taken up by increased stocks of squid and fish, though there are no data adequate to show this. It would be surprising though if squid and fish had not responded in the same way as seals and

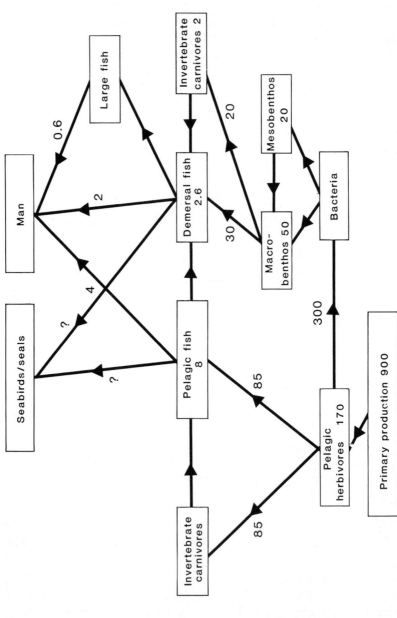

Figure 5.14 North Sea food web based on major groups of organisms and inserting values for estimated annual production. (Based on Steele, 1974.)

seabirds. The indications are that seabirds in the Antarctic have responded faster to the reduction of whale stocks than have seals, and this might be predicted since seabirds (here we are generally considering penguins since these make up more than 80% of Antarctic seabird biomass) tend to have a shorter generation time and higher fecundity than seals and so a greater capacity for increase.

These calculations also show that krill consumption by seabirds is now greater than that by whales, and nearly as much as taken by seals. Thus, in the present day Southern Ocean ecosystem, seabirds and seals are more important in terms of energy flow than are the remaining whale stocks. Unfortunately we cannot assess the relative importance of fish and squid, but any consideration of Southern Ocean ecology must take seabirds into account as one of the principal consumers. Harvesting of krill by man would clearly result in a reduction in availability of krill to seabirds as well as seals and whales. It is difficult to guess how these different consumer populations would be affected, but it seems likely that species with shorter life expectations would be the first to decline in numbers. Seabirds with limited foraging ranges (e.g. penguins), specialised feeding methods or a high dependence on krill would probably be most severely affected. It is also worth remembering that the Southern Ocean ecosystem is already considerably distorted from its natural balance by the effects of whaling, so that seabird populations are now artificially high.

5.5.3 *The North Sea*

The North Sea provides one of the richest yields of fish in the world. Detailed data on catches, fishing effort, and stock size and composition have been collected for commercial fish species since the creation of the International Council for the Exploration of the Seas (ICES) in 1899. The long history of its exploitation is well recorded (Hempel, 1978). The structure and productivity of the North Sea marine ecosystem has also been investigated in detail (Steele, 1974). Primary production, although patchy on a small scale, is relatively uniform over the North Sea as a whole. The major consumers of zooplankton are fish; large pelagic shoaling species (herring and mackerel) or generally smaller fish that often are partly bottom living and rise to feed on zooplankton primarily at night (sand-eels, sprats and Norway pout). The small fish are consumed by seabirds, seals and by large predatory fish such as mature cod and whiting (Figure 5.14). As a result, seabirds and predatory fish may potentially compete for small fish prey.

The British Isles hold large breeding populations of seabirds. Changes in the sizes of these seabird populations have been documented for several species in many colonies and provide an opportunity to examine correlations between changes in seabird and fish populations over a long period of time. In this section we

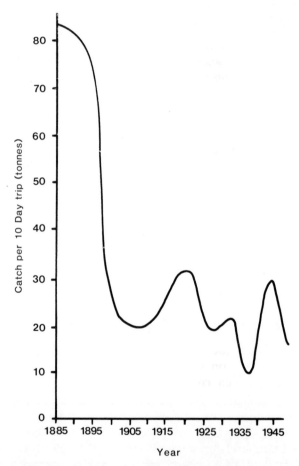

Figure 5.15 Catch per unit effort (i.e. index of population size) for all demersal fish in part of the North Sea from 1885 to 1947. (From Lundbeck, 1962.)

shall outline the changes in fish stocks in the North Sea since the end of the nineteenth century and consider the responses shown to these changes by seabird populations.

The introduction of steam trawlers and power winches at the end of the nineteenth century, together with the development of the otter trawl, greatly increased fishing power. Lundbeck (1962) showed that a severe reduction in whitefish biomass occurred almost immediately (Figure 5.15), particularly in the southern North Sea where fishing was then most intense. He attributed this change to overfishing as a result of increased fishing effort, and this interpretation is strengthened by the tendency for the stock to show temporary increases during the two world wars, 1914–18 and 1939–45, when fishing effort was much reduced. However, demersal fish species are well known for their habit of producing occasional 'bumper' years of high recruitment, possibly as a result of a fortuitous coincidence between the timing of the larval stage and peak zooplankton abundance (Cushing, 1975). The high peak of whitefish biomass in the 1880s could have resulted from such an event, though if this is the explanation it has never been repeated since on such a scale. A more likely explanation is that fishing effort since 1900 has been so high that whitefish stocks

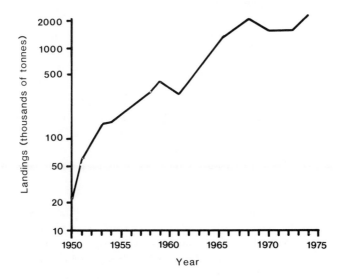

Figure 5.16 Total annual landings of industrial fisheries (predominantly immature herring and mackerel, sprats, sand-eels and Norway pout) in the North Sea. (From Hempel, 1978.)

have remained at a much lower level than would be the case in the absence of fishing. Whitefish have responded to the improved food supply as a result of stock depletion. Haddock and whiting growth rates increased so that both species reached reproductive condition at an earlier age (Jones and Hislop, 1978; Daan, 1975). Cod showed no change in growth rate, but the age at maturity decreased, so that cod reached reproductive age at a smaller size (Daan, 1978).

Depletion of herring and mackerel stocks in the North Sea had begun before 1950, but was greatly accelerated by the introduction of purse-seining. In a few years of fishing on immature fish for reduction to fish-meal and oils, stocks were reduced by 90%. The fisheries for adult herring and mackerel before 1960 will have been directly beneficial to seabirds by reducing the average size of fish in the populations without greatly reducing stock biomass, so that a higher proportion of the stock will have been in the size range suitable for seabird consumption. Andersen and Ursin (1977) found that the reduction in stocks of herring and mackerel is likely to have led to increases in the populations of their ecological competitors, particularly sand-eels. Evidence that such increases have taken place is not readily available as sand-eels were of no commercial interest until relatively recently. However, there is good evidence for a large increase in sand-eel stocks in both the northeast and northwest Atlantic (Sherman *et al.*, 1981). Andersen and Ursin's model also predicted that whitefish stocks would increase as a result of decreased predation on their larvae by herring and particularly by mackerel. Such increases have occurred, and would be very difficult to explain except in terms of such an ecosystem interaction (Hempel, 1978). As a corollary of their recovery, the growth rates of whitefish have fallen again, which suggests that the superabundance of food generated by their stock depletion no longer exists, although whitefish stock recovery has not approached more than a small fraction of the apparent abundance of the 1880s.

The reduction of herring and mackerel stocks will have been beneficial to seabirds in so far as it allowed an increase in their main food during the breeding season, sand-eels, although partial recovery of whitefish stocks will have increased predation on adult sand-eels. Also, there is evidence that seabirds feed on other prey outside the breeding season, when sand-eels tend to spend most of the time buried in the sea bottom, and are probably unavailable to seabirds. Depletion of herring and sprat stocks may be of great

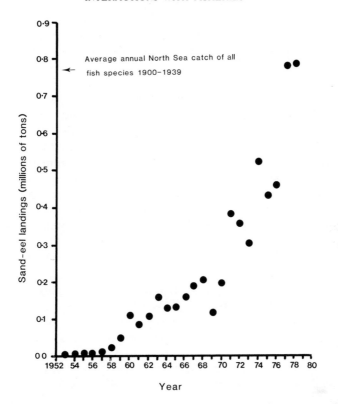

Figure 5.17 Annual landings of sand-eels caught in the North Sea. (From Furness, 1982.)

importance to the winter feeding and survival of seabirds. A result of these ecosystem changes has been the development of industrial fishing, particularly for sand-eels and Norway pout. In the 1970s the total landings of industrial fisheries in the North Sea exceeded landings of whitefish, averaging over two million tonnes per year (Figure 5.16). Sand-eel catches have increased most dramatically in the 1970s (Figure 5.17) although little is known of the maximum sustainable yield for this species. Around Shetland, the sand-eel fishery catches mainly one- and two-year-old fish between March and May, then mainly the first year (O-group) fish as they recruit into the fishery through rapid growth in June to August. Most Shetland seabirds feed on the one- and two-year-old sand-eels, which form the bulk of the diet of Shetland seabird chicks, and there is clear concern that intense fishing on these earlier in the year may result in shortages for the seabirds. There is little evidence of this as yet.

The total annual landings of whitefish from the North Sea remained around 0.5 million tonnes from 1910 to 1960, then increased slightly to over one million tonnes between 1965 and 1975 as a consequence of enhanced recruitment, which may have been a response to decreased predation on whitefish larvae. Subsequently, whitefish stocks have declined again, as have catches, and this may reflect environmental changes, effects of fishing, or some increase in predation on white-fish larvae. Legal restrictions on the minimum sizes of fish that can be marketed, and fluctuations in market demand and catch quotas, result in a proportion of each catch being discarded, mainly dead, into the sea. This proportion has increased with the reduction in average fish size; now a higher proportion of the population is caught as soon as it is recruited into the fishable stock. About 30% of a catch of whitefish may routinely be discarded at sea (Jermyn and Hall, 1978). Shellfish trawlers may discard large quantities of even smaller fish that are retained by their fine mesh nets, particularly when catching Norway lobsters. Offal, which is probably about equal in weight to the amount of discards from whitefish boats (Bailey and Hislop, 1978), and discarded whole fish, provide a source of food for Fulmars, large gulls, Great Skuas, Gannets and Kittiwakes, that would otherwise be unavailable to seabirds. The yield of whitefish to commercial fisheries in the North Sea could be increased in the long term by increasing minimum net-mesh size, allowing smaller fish to escape and continue their growth until capture at a later date. Offal may be increasingly used in future to mix with industrial catches, while discarding would decrease if mesh sizes were increased. Such changes would clearly reduce food availability to scavenging seabirds.

There is a clear dominance hierarchy among seabirds feeding at fishing boats. Fulmars displace all other species, and tend to select offal, rarely attempting to swallow or tear open whole fish. Gannets and Great Skuas dominate gulls, and obtain first choice of the discards. Great Black-backed Gulls dominate Herring Gulls, and Kittiwakes are only able to feed on scraps that the other species ignore. Great Skuas in Shetland consumed up to 57% of the whitefish discarded by the fishing industry, depending on the availability of sand-eels, their preferred prey (Furness and Hislop, 1978). The hierarchical arrangement suggests that when sand-eels are scarce and skuas take a disproportionate share of the available discards, gulls are likely to

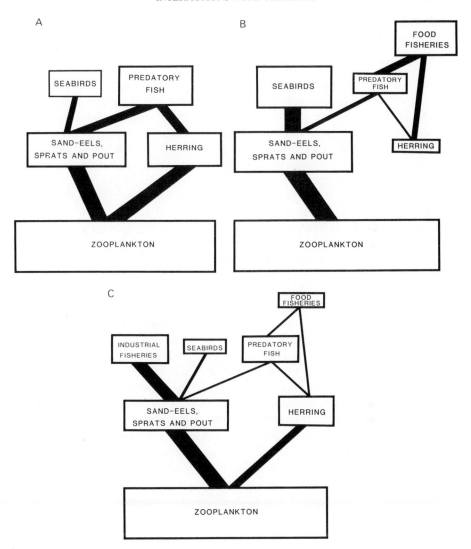

Figure 5.18 A diagrammatic representation of energy flow through parts of the North Sea ecosystem: (*a*) in the late 19th century, (*b*) in the 1960s when herring stocks were depleted, (*c*) possible picture in the 1990s with increasing industrial fishing, conservation of herring and whitefish stocks and possibly a decline in seabird populations as a result. Sizes of rectangles and widths of flow lines are intended to be indicative of magnitude although only for comparisons between stages *a*, *b* and *c*.

turn to other feeding opportunities. Interspecific relationships may play an important role in the responses of seabird populations to changing fish stocks or fisheries practices.

Seabirds have benefited from two quite distinct effects of North Sea fisheries. Firstly, the provision of discards and offal has increased food available to scavenging species. Secondly, and of greater overall importance, ecosystem changes leading to an expansion of sand-eel stocks has increased food availability to almost all seabirds. These changes are summarized diagrammatically in Figure 5.18. Although numbers of the alcids are thought to have declined at many colonies, most seabird populations in Britain and Ireland increased over the first 80 years of the 20th century. Several are now showing signs of reaching a population ceiling, or possibly a temporary check. Rates of increase have been highest in the gulls, Great Skua and Fulmar, species which make use of discards and offal as well as sand-eels, and for many species have been greater in Shetland and east Britain than in west or south Britain. Some of these changes may be partly due to reductions in human persecution, but it seems likely that food changes have been important. Unfortunately we do not understand how these seabird populations are regulated, so it is impossible to come to firm conclusions regarding the role of food, or whether food supplies during the breeding season are more or less important than those in winter. From data collected over a period of 31 years at a Kittiwake colony in NE England there is reason to believe that many changes at this colony in numbers of breeding pairs, the survival rate of adults, timing of breeding and breeding success can be related to changes in the availability of fish during the very early part of the breeding season when nest sites are reoccupied (Coulson and Thomas, 1985b). So far, no other studies have been undertaken to determine when in the year food supplies are most critical for seabirds at other colonies.

5.6 Conclusions

It is clear from studies reviewed in this chapter that seabirds are important predators of lower trophic levels in marine ecosystems. Their numbers are affected by human fisheries. In some cases exploitation of competitors (whales or predatory fish) has been beneficial to seabirds, while in others (e.g. overfishing of Peruvian anchovies and South African sardines) it has been detrimental. Whether or not predation on fish stocks by seabirds has any significant effect on yields to man is less clear. This is only likely where seabirds and commercial fisheries exploit the same age groups of the same species of fish, as in fisheries for anchovies and sardines. The dramatic declines in seabird numbers in Peru, South Africa and Namibia as a consequence of intensive fishing for anchovies and sardines suggest that seabirds are very sensitive to a reduction in the density of their prey, a conclusion also reached as a result of theoretical modelling by Ford *et al.* (1982). In this case, the interaction between seabirds and fisheries appears to be largely one-way. Seabird populations are vulnerable to changes in their food stocks induced by fishing, but apparently, have little effect on the yield to man.

CHAPTER SIX

MONITORING MARINE ENVIRONMENTS

6.1 Introduction

The oceans are the sink for many of man's waste products and substances lost by accident. These quickly become enormously diluted, so that ocean dumping or discharging into estuaries or coasts provides a cheap and apparently safe method of disposal. Two factors can alter this. In some areas, polluted waters mix little with general ocean circulations. Pollutants in the relatively enclosed Baltic Sea and Mediterranean Sea can build up more rapidly than they would in the North Atlantic Ocean, causing local pollution problems. Secondly, a wide range of substances can be taken up by organisms and concentrated to many times the level found in the surrounding water (bioaccumulation). Seaweeds and marine invertebrates of many groups can concentrate some metals by factors of 10 000 to 100 000 times that found in seawater (Bryan, 1984). Further increases in concentration can occur as a result of biomagnification. That is, animals higher up the food chain may accumulate higher concentrations than found in their prey as a result of feeding on many prey organisms. Although biomagnification does not always occur, since higher animals can regulate levels of many substances in their bodies, top predators can often be exposed to pollutants several orders of magnitude more concentrated than might be expected from information on the rates of pollutant input into the sea.

Because seabirds tend to feed in the upper trophic levels of marine food webs and are both numerous and conspicuous, they provide an excellent monitor of the health of marine environments. They are being used to monitor organochlorines, heavy metals, oil and plastics throughout the world's oceans, and it has been suggested that seabirds may also provide indications of changes in fish stocks in a more immediate and economical way than can be done by fishery research vessels.

6.2 Plastics

The world's seas and oceans are widely contaminated with plastic flotsam and jetsam. Since plastic is not readily broken down, and the production of plastics is increasing worldwide, levels of contamination are likely to increase. Plastic fragments may be ingested by fish, seabirds or marine mammals, and their effects on the digestive system or physiology are unknown. Plastics can be categorized into two groups. User-plastics are fragments of items such as polystyrene cups, plastic toys or torn pieces of synthetic netting. The other group, plastic pellets, comprises the raw material from which user products are moulded. Plastic particles, which are largely the raw plastic pellets but contain a small quantity of eroded fragments from consumer items, are present at average densities of 1000 to 4000 per square kilometre on the surface of the North Atlantic, South Atlantic and Pacific Oceans and are probably now distributed over the other oceans and seas as well (Morris, 1980).

Plastic particles have been found in the gizzard or proventriculus of many species of seabirds from Alaska to the Antarctic. It has been suggested that seabirds mistake plastic particles for zooplankton or fish eggs, but it is also possible that some are taken in inside the stomachs of prey animals eaten by seabirds. It is known that polychlorinated biphenyls, or PCBs (see 6.4) are adsorbed on to the surface of plastic, and so ingestion by seabirds may increase their uptake of PCBs. There are a few instances of plastic causing ulceration of the stomach of seabirds, but the most likely physiological damage caused by ingesting plastic is probably the reduction in functional volume of the gizzard, leading to a reduction in digestive capability (Connors and Smith, 1982). Connors and Smith reported a negative correlation between the number of plastic particles in the gizzard of Grey Phalaropes and their body mass. This correlation failed to reach statistical significance, partly because only a very small number of birds had been examined, but it suggested the need for further research into the occurrence and effects of plastic in seabirds.

Charadriiform and sphenisciform seabirds (such as gulls and penguins) generally have few plastic particles in the gizzard, and this appears to be due to their habit of regularly regurgitating pellets of indigestible matter. Gulls and skuas generally regurgitate at least one pellet per day, and these occasionally contain plastic particles.

Procellariiform seabirds generally have the most plastic in the gizzard, although there is great variation between individual birds, with even some individuals of what are generally the most contaminated species containing no particles (Furness, 1985a, b). The extent to which procellariiforms can eject gizzard contents is unknown. Gizzard and proventriculus anatomy differ from that of charadriiforms (Figure 6.1), and the constriction at the entrance to the gizzard may make regurgitation of gizzard contents difficult. Albatrosses certainly can eject pellets of squid beaks, but there is no record of shearwaters or petrels doing this. A plastic telemetry transmitter fed to a Laysan Albatross remained in the stomach for over 40 days (Pettit *et al.*, 1981) and squid beaks fed to a Shy Albatross remained in the stomach for at least 6 weeks (B.L. Furness, 1983). From consideration of the state of erosion of particles in Short-tailed Shearwaters, Day (1980) suggested that these may have a residence time of about 15 months, though this estimate could do with verification by direct

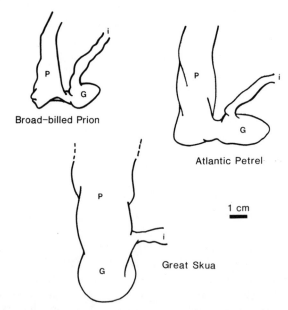

Figure 6.1 Drawings of the external structure of the gizzard and proventriculus dissected from Broad-billed Prions, Atlantic Petrels and Tristan Great Skuas. *P*, proventriculus, *G*, gizzard, *i*, intestine. (From Furness, 1985.)

testing. Plastic particles may be worn down in size until they can pass through the intestine, but examination of the distribution of sizes of particles in different species does not strongly support this idea. Although in most cases larger seabirds clearly select larger particles, they also contain a variety of small and very small particles, with no evidence of a cut-off point as small particles are lost (Figure 6.2). Both northern and southern hemisphere procellariiforms are heavily contaminated with plastic (Table 6.1) and differences between species can probably be attributed to different feeding behaviours. When removed from a dead bird, the gizzard can be inflated to about four times its relaxed volume, but this may not be possible *in vivo*. However, the volume of plastic may represent a large part of the space available for food in the most contaminated birds, so that a reduction in digestive ability may be likely. Within a seabird species, body mass is closely related to body size, so that variation due to size should be taken into account before considering effects of ingested plastic. Multiple regressions relating mass to the best measures of body size and the quantity of plastic in the gizzard suggest that plastic contamination may reduce body mass (presumably by causing a loss of stored energy reserves), but that the effect is rather difficult to demonstrate statistically. The ubiquitous nature of plastic particle pollution of oceans, and increasing use of plastics, means that there is an urgent need for experimental studies to determine the residence time and physiological effects of ingested plastic in seabirds.

6.3 Oil

Although there is limited seepage of crude oils from sea-floor rocks into the sea, oil pollution usually results from washing of oil-tanker tanks at sea or small accidental spillages. Major disasters such as tanker accidents or oil-well blow-outs are rare. Seabirds are vulnerable to oil in a number of ways. Oil soaks the plumage, destroying buoyancy and insulation. Oil ingested when birds attempt to preen the plumage clean may lead to gastro-intestinal irritation, damage to liver and kidney function, and general shock. Adult seabirds lightly oiled may transfer oil on to eggs, leading to an increased risk of the suffocation of the embryo.

Major accidents, such as the grounding of the oil tanker *Torrey Canyon* in 1967 off the Cornish coast, can cause spectacular loss of

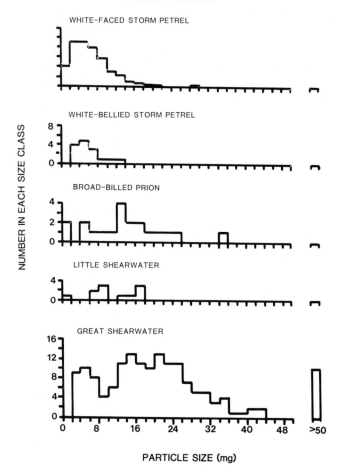

Figure 6.2 Size (mass) of plastic particles ingested by seabirds at Gough Island, South Atlantic Ocean. (From Furness, 1985.)

life among seabirds and lead to an increased awareness among the general public of this problem. However, analysis of ring recoveries of oiled seabirds shows clearly that most of the mortality due to oil is caused by frequent but small discharges, forming chronic oil pollution (Baillie, 1987). The *Torrey Canyon* incident killed well over 10 000 auks, and numbers of auks breeding at the nearby Sept Isles colony in Brittany were reduced within that year by about 80% (Bourne, 1976), indicating that oil pollution can be a serious additional mortality factor for adult seabirds.

Table 6.1 Contamination of seabirds by plastic particles. (Data from Furness, 1985a, b.)

Species	Locality	Number examined	Mean number of pellets per bird	Maximum pellets per bird	Maximum percent of relaxed gizzard volume taken up by plastic
British Storm Petrel	St Kilda	21	0.0	0	0
Leach's Petrel	St Kilda	17	2.9	12	33
Manx Shearwater	Rhum	10	0.4	2	3
Fulmar	St Kilda	8	3.9	16	24
Fulmar	Foula	13	10.6	40	59
Great Shearwater	Gough Is.	13	12.2	53	81
White-faced Storm Petrel	Gough Is.	19	11.7	40	96
Broad-billed Prion	Gough Is.	31	0.6	3	14
Little Shearwater	Gough Is.	13	0.8	11	23
Common Diving Petrel	Gough Is.	19	0.0	0	0

Oil on the sea surface loses lighter hydrocarbons by evaporation, until eventually only rather inert tarballs remain. Evaporation is temperature-related. Oil slicks remain for longest in cold conditions. For this reason, seabirds are oiled in greatest numbers in winter. Surveys of dead seabirds on beaches show that auks are most vulnerable (Table 6.2), while the proportion oiled is highest in areas of greatest shipping traffic such as the English Channel. Details of oiling incidents and the regional, seasonal and species patterns of oiling have been reviewed many times (e.g. Bourne, 1976; Stowe and Underwood, 1984). In the United Kingdom between 1966 and 1983 there were 144 reported oiling incidents, each causing the death of more than 50 seabirds (Stowe and Underwood, 1984). About 70% of the 89 213 birds involved were auks, principally Guillemots. Most incidents occurred between November and March. Patterns in the rest of Europe are similar, but have often involved larger numbers of casualties, particularly of seaducks off Denmark (Stowe and Underwood, 1984).

Since most oiling of seabirds occurs at sea and in winter, understanding the distribution of seabirds at sea outside the breeding season is of great importance. As a result, an attempt has been made to map the distributions of seabirds, particularly species vulnerable to oil, over the North Sea at all times of year (Blake *et al.*, 1984). Such data could, in theory, provide a measure of the numbers of

Table 6.2 Species composition of 49 079 casualties in 119 oil pollution incidents in the United Kingdom, July 1971–June 1983 (from Stowe and Underwood, 1984)

Species or group	Casualties	Percent of total
Guillemot	26 386	53.8
Razorbill	3480	7.1
Seaducks	2628	5.4
Gulls	1453	3.0
Shag or Cormorant	1024	2.1
Black Guillemot	743	1.5
Gannet	635	1.3
Grebes	445	0.9
Puffin	398	0.8
Waders	353	0.7
Divers	347	0.7
Fulmar	235	0.5
Other wildfowl	742	1.5
Others	6156	12.5
(All auks	35 061	71.4)

birds at risk in the event of an oilspill anywhere in the North Sea at any time of year. However, the high mobility of seabirds and their characteristically patchy and unpredictable distribution patterns make this very difficult to achieve.

Regular surveys of birds washed up on beaches provide a useful indication of the changes in mortality of seabirds from year to year. Large-scale ringing of auks and Shags also provides evidence of the scale of oil-related mortality and the origins of populations involved in kills (Baillie, 1987).

Only prevention of spillage is an adequate control of oil pollution. Deterring seabirds from entering areas contaminated by oil is unlikely to be viable, and rehabilitation of cleaned birds has proved to have very limited success with most species. Although some 8000 seabirds were cleaned after the *Torrey Canyon* disaster, only a handful survived to be released back into the wild, and probably none of these survived to breed (Clark, 1978). The only seabirds that can be rehabilitated after oiling with a fair degree of success are penguins. Of 5565 Jackass Penguins collected oiled, 3030 (54%) were subsequently released in good condition. Following seven major incidents when 4639 penguins were collected oiled, 788 individuals (17%) were resighted after release, in the vicinity of breeding colonies (Morant *et al.*, 1981). Some of these are known to have bred successfully, and in view of the endangered nature of the species, this rehabilitation programme seems to be very worthwhile. However, the breeding success of rehabilitated penguins has not been compared with that of normal birds.

With the development of oilfields in deep waters close to major seabird colonies (such as the oilfields to the west of Shetland), there is an increased concern about the impact an oil spill would have on these populations during the breeding season, when most of the population is concentrated in a relatively small area. Ford *et al.* (1982) developed a simulation model to estimate the responses of colonial seabirds (Guillemot and Kittiwake populations were simulated in the model) to oil spills within the foraging range during the breeding season. They looked both at effects of adult mortality due to oiling and at effects of reduction in the available foraging area (sea not covered by oil within the foraging range). The latter effect could cause a dramatic decrease in breeding success if assumptions about the relationships between foraging range, food provisioning rate, chick growth rate and survival are correct.

A major conclusion of this model, which is the first to tackle this question, is that many critical aspects of seabird breeding biology are not sufficiently understood, and require further study before this kind of modelling approach can be used to its full potential in a predictive sense. However, it is a major strength of the modelling approach adopted that it displays clearly the areas of ignorance and thereby the data of critical importance that need to be collected.

6.4 Organochlorines

Chlorinated hydrocarbon pesticides, such as DDT, were introduced in the late 1940s, but it was not until the mid-1960s that it was realized that these had become widely distributed in marine environments and accumulated in seabirds even in the Antarctic. Concentrations of organochlorines are measured by quantifying peaks separated by gas-liquid chromatography. DDT is rapidly broken down into DDD and then DDE, but DDE is very stable and is the form of the insecticide that accumulates in wildlife. Several peaks are produced on the chromatogram representing the small amounts of DDT and DDD and, usually, larger amounts of DDE. Peaks are also produced by other compounds, such as HEOD, the constituent of the insecticide dieldrin. This is similarly accumulated by wildlife. PCBs (polychlorinated biphenyls) were introduced in the 1930s for numerous industrial uses, particularly in paints and plastics, and PCBs were also found in samples of seabird tissues being analysed for organochlorine insecticides. All these compounds have fairly similar toxic effects on seabirds (and other vertebrates). They are fat-soluble, and levels of each tend to correlate between seabirds, so that it is difficult to attribute harmful effects to any one of these contaminants. There is an enormous literature on the occurrence and effects of organochlorines in seabirds, with many general reviews on the subject, for example Bourne (1976); Anon (1983).

6.4.1 Effects of organochlorines

All of the chlorinated hydrocarbon compounds appear to be able to damage the nervous system and liver of vertebrates. When deposited in body fat, organochlorines are inert and harmless, but mobilization of fat releases the organochlorines into circulation in the blood. In addition to damaging the liver and nervous function, circulating organochlorines tend to depress the appetite of birds, which can have a positive feedback effect resulting in further mobilization of lipids and greater damage by organochlorines.

(a) *Hormonal interactions.* DDT, its derivatives, and PCBs are similar in molecular shape to molecules of the hormone oestrogen, and it has been shown that these organochlorines show oestrogenic activity when tested in assays (Bitman and Cecil, 1970). PCBs can also affect the function of the thyroid gland. Experimental dosing of Guillemots and Lesser Black-backed Gulls showed that PCBs affected the pituitary, causing a dose-related decrease in the production of thyrotrophin, and so causing hypothyroidism (Jefferies and Parslow, 1976). Interestingly, Guillemots were much more sensitive to the PCB doses than were the gulls, and the range of PCB concentrations in the livers of these experimental birds was similar to that found in wild seabirds collected from various localities in the North Atlantic. This last point suggests that prevailing levels of PCBs may be sufficient to have harmful sublethal effects on at least the more sensitive species of seabirds.

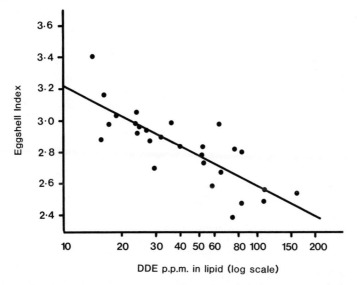

Figure 6.3 Eggshell thickness index in relation to concentration of DDE in egg lipid for British Gannets. (From Parslow and Jefferies, 1977.)

(b) *Eggshell thinning*. Eggshell thinning induced by chlorinated hy-drocarbons has been a major factor in the population declines of some predatory and piscivorous birds (Ratcliffe, 1970). Studies of birds both in the wild and in the laboratory have shown that DDE is the main cause of eggshell thinning, through its direct influence on calcium metabolism. Dieldrin (HEOD) is far less effective, while PCBs may not cause shell thinning at all. Shell thinning has been demonstrated in many seabirds, for example British Gannets (Figure 6.3), British Shags (Ratcliffe, 1970), Brown Pelicans (Blus *et al.*, 1974 and references therein), Guillemots (Gress *et al.*, 1971; Dyck and Kraul, 1984), and Ashy Storm Petrels (Coulter and Risebrough, 1973). However, eggshells in Great Lakes Herring Gulls that had extremely high organochlorine levels and low hatching success were found to be only about 2% to 8% thinner than normal (Gilman *et al.*, 1977) while thinning of at least 15% is thought to be necessary before increased eggshell breakage occurs in the wild (Blus *et al.*, 1974). Great Skuas also accumulate high levels of organochlorines, but show no detectable eggshell thinning (Furness and Hutton, 1979). An indication of the differences in sensitivity of different seabirds is given in Table 6.3. Gulls appear to be much less sensitive in this respect than Brown Pelicans, Gannets, Guillemots or cormorants.

Dyck and Kraul (1984), looking at pollutants and eggshell thinning in Guillemots from the Baltic, found that eggshell thickness, pollutant levels and seawater salinity were all intercorrelated in a complex way, making it difficult to determine cause and effect. Shell thick-ness declined with increasing concentration of methyl-mercury, and

Table 6.3 Eggshell thinning in some seabirds in relation to levels of DDE in the egg lipid.

Species	Locality	DDE in lipid (ppm)	Shell thickness (% reduction)
Herring Gull	Wisconsin, USA	2525	19%
Brown Pelican	California, USA	2250	53%
Brown Pelican	California, USA	1225	34%
Herring Gull	Wisconsin, USA	972	10%
Guillemot	Baltic	450	9%
Guillemot	California, USA	300	12%
Double-crested Cormorant	USA	208	8%
Gannet	Scotland	150	20%
Great Skua	Iceland	70	0%
Common Tern	Canada	50	0%
Common Tern	Canada	30	4%
Great Skua	Shetland	30	0%

increased with seawater salinity. However, methyl-mercury levels in eggs also declined with increasing salinity. No relationship could be shown between shell thickness and DDE levels, but this could be attributed to the low DDE levels in most eggs. No other field studies and no laboratory studies indicate that mercury in any chemical form causes eggshell thinning, so Dyck and Kraul (1984) suggest that seawater salinity may be the causal factor. Surplus salt is excreted by the salt gland, and this is mediated by the enzyme carboanhydrase, which also plays an important role in the shell gland. The interaction between the activities of the two glands may result in the observed correlation between salinity and shell thickness. This is not to say that DDE effects are any less important than previously thought, but simply that effects of salinity may obscure some of the relationships between DDE levels and shell thickness.

(c) *Reduced breeding success.* Hens fed diets containing PCB produced normal quantities and sizes of eggs, with normal thickness of shells, but the hatchability of the eggs was reduced by PCB. Other organochlorines have much less effect on egg hatchability. Egg hatchability, chick survival and nest site defence were all lowest at Herring Gull colonies most contaminated with PCBs (Table 6.4). By hatching eggs in an incubator, it was possible to show that part of this effect was due to a lower hatchability of the eggs from colonies with high organochlorine levels (Table 6.4) and part due to effects on adult behaviour.

Gilman *et al.* (1978) then injected organochlorines into Herring Gull eggs in the wild and monitored hatchability. Developing embryos took up the injected organochlorines. However, concentrations of up to 500 ppm of organochlorines had no detectable effect on embryo survival (though the solvent in which the organochlorines were injected did drastically reduce embryo survival in both experimental and control eggs). These results suggest that the reduced viability is due not to the direct toxicity of the organochlorine to the embryo, but to effects of the organochlorine on the female, affecting the quality of the egg she produces in some way.

Blus *et al.* (1974) used a slightly different approach to examine the influence of organochlorines on Brown Pelican reproductive success. One egg was taken from each of 93 marked nests at a colony suffering reduced breeding success, and organochlorine levels were

Table 6.4 PCB levels in eggs, hatching and fledging success in Herring Gull populations on the Great Lakes and hatching success of eggs in artificial incubators (from Gilman et al., 1977)

Colony	PCB levels in eggs (ppm wet wt)	Hatching success (%)	Embryo death (%)	Fledging success (%)	Fledglings per pair	Hatching success in artificial incubator (%)
Lake Ontario	142	19	35	25	0.15	25
Lake Erie	66	63	17	73	1.41	64
Lake Superior	60	80	9	58	1.38	—
Lake Huron	52	72	6	66	1.48	57

Table 6.5 Organochlorine residues in Brown Pelican eggs from successful and unsuccessful nests (from Blus *et al.*, 1974)

Nest status	Geometric mean organochlorine concentration (ppm wet weight)				
	DDE	DDT	DDD	Dieldrin	PCB
Successful	1.8	0.1	0.3	0.3	5.5
Unsuccessful	3.2	0.2	0.5	0.5	7.9
Significance of difference	$p < 0.005$	ns	$p < 0.025$	$p < 0.025$	ns

determined. Hatching success of the other eggs in each clutch was then monitored and correlated with organochlorine levels (assuming that intra-clutch variation in organochlorine levels was negligible compared to variation between clutches). Levels of different organochlorines were correlated, making it difficult to identify which caused nest failure, but both DDE and dieldrin levels were significantly higher in eggs from unsuccessful nests. Levels of DDD, DDT and PCBs were not significantly related to nest success, though they tended to also be higher in eggs from unsuccessful nests (Table 6.5). In this species, levels of DDE were about half the levels of PCBs. In most seabirds studied the ratio of PCBs to DDE is greater than found in Brown Pelicans, so the lack of a detectable effect of PCBs may be due to its relatively low level by comparison with DDE. Alternatively, Brown Pelicans may be sensitive to DDE but relatively insensitive to PCBs.

Harris and Osborn (1981) performed a careful field experiment to examine the influence of PCBs on reproductive success of Puffins. Samples of adult breeding Puffins were implanted with 30–35 mg PCB, or with sucrose (a control group). The survival and breeding of these birds was then followed, and some individuals were later killed for PCB analysis. PCB body burden in dosed Puffins reached an average of 6 mg PCB, ten times the amount in control birds and more than found in Guillemots in the 1969 Irish Sea seabird wreck. However, most of this was stored in body fat reserves. Levels of PCB in the liver of dosed birds rose to between 2 and 48 ppm (wet weight) during the first year after implantation, while controls contained negligible amounts in the liver. However, the dosed birds showed no changes whatsoever in breeding performance or behaviour. Harris and Osborn (1981) review the literature on effects of

PCBs and conclude that their experiment failed to show the harmful effects of sublethal doses of PCBs in Puffins, that might have been expected. Various possible explanations for this may be put forward. PCBs might only be toxic in conjunction with DDE. Puffins might be unusually insensitive to toxic effects of PCBs or may be unusually capable of detoxifying PCBs. Possibly more importantly, this experiment was carried out at the Isle of May, east Scotland, where Puffin numbers were rapidly increasing, and the birds had high body fat reserves. Probably they had little need to draw on these reserves, so that PCBs would not have reached target organs in dangerous quantities. Levels in the liver, kidney and muscles of dosed birds were not high when compared to levels found in many undosed seabirds in other areas. Harris and Osborn point out that the conclusions of their study might have been different if it had been performed at a colony where Puffins had to mobilize fat reserves to cope with times of food shortage or stress. This interpretation is strengthened by the fact that there was no correlation between the levels of PCBs in the egg and the body fat of individual dosed female Puffins (Harris and Osborn, 1981). This suggests that the lipids entering the egg were largely derived from food rather than requiring the female to draw on her body reserves, and this would be more likely to occur if feeding conditions were highly favourable, since correlations between organochlorine levels in eggs and the laying female have been found in other studies.

6.4.2 Monitoring of organochlorines

Organochlorine levels in seabirds are normally determined in lipid extracted from eggs, or from samples of liver, muscle or body fat taken from adults. Many analyses have been performed on birds found dead or even found dying in convulsions. Not surprisingly, concentrations in the liver of such birds tend to be extremely high, since body fat reserves have been used up. However, the total body burden may be quite small, since mobilization of fat reserves removes the main store of organochlorines and causes the high liver levels. Nearly half of the determinations of organochlorines in 250 seabirds listed in Bourne (1976) were samples from dead or dying individuals, and in many cases the organochlorine levels tabulated include mixtures of healthy and starved birds, with no indication of total body burdens. Such analyses tell us rather little about organochlorine pollution of the environment, and can give a misleading impression of the variation in levels of contamination within a population.

The distribution of organochlorines between organs varies greatly from bird to bird and it may be that this information could be of use in the assessment of the pollution condition of the bird, but at present we do not know enough about the reasons for tissue distribution patterns observed, except that healthy seabirds have most of the residues in fat reserves, and high levels in the liver and other organs tend to imply poor condition.

Robinson *et al.* (1967) showed that young seabirds reach a dynamic equilibrium of organochlorine levels within a short period of exposure; within the first year of life in Shags and Kittiwakes in east Britain. One of their main foods, sand-eels, showed a pronounced seasonal cycle in organochlorine levels, suggesting that the residence time of organochlorines in animal tissues is relatively short. That study, and that of Tanabe *et al.* (1984) showed that organochlorine levels were higher at higher trophic levels, and the latter study showed that the more lipophilic and less metabolizable organochlorines tended to represent a higher proportion of the total in animals at higher trophic levels. This indicates that much of the organochlorine burden is broken down and lost, but that this is a selective process. Seabirds therefore provide a slightly different measure of organochlorine pollutants than would samples of seawater or plankton, but they have the advantage of concentrating the pollutants into more easily measured amounts and they average out short-term and small-scale geographic variation. They therefore provide a better general measure of pollution levels for an area of sea or ocean.

Seabird eggs provide a more straightforward index of organochlorine levels and also avoid the need to kill healthy birds. Levels in eggs usually, but not always, correlate with levels in the birds laying them. However, Mineau (1982) found that organochlorine levels increased with laying sequence in Herring Gulls. This is probably due to an increased use of fat reserves by the female as successive eggs are formed. Body fat will contain more organochlorine residues than lipid derived directly from food. It may be a common phenomenon in seabird clutches, but the difference between first-, second- and third-laid eggs reported by Mineau was small compared to differences found between birds or between populations, so will not seriously affect the value of eggs as indicators of organochlorine levels in seabird populations, and hence in marine ecosystems.

Monitoring of organochlorine levels has shown changes over time which can be attributed to changes in the use of these substances. Shags were chosen by Coulson *et al.* (1972) as a good indicator species because the organochlorine levels were known to be high enough to allow changes to be measured, the birds normally remain within 60 km of the natal locality so that local pollution levels can be monitored, the diet was known, it had already been shown that after 6 months of age organochlorine levels were independent of age in that species, and egg levels correlated with adult levels. Dieldrin and DDE levels were monitored in samples of 9 to 69 eggs each year from 1964 to 1971 (Coulson *et al.*, 1972) and showed that restrictions in the use of aldrin and dieldrin in the mid to late 1960s were reflected by reductions in levels of HEOD in the Shag eggs (Figure 6.4). DDE levels also fell after reaching a peak in 1967–68, possibly also as a consequence of reduced use of this insecticide.

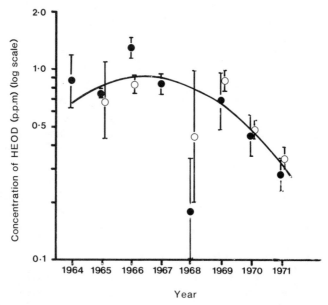

Figure 6.4 Concentrations of dieldrin in Shag eggs from 1964 to 1971. Mean (geometric) annual concentrations and confidence limits are indicated for Farne Islands (●) and Isle of May (○). The line represents the best-fit quadratic relationship. (From Coulson *et al.*, 1972.)

Fimreite *et al.* (1977) found no detectable decrease in DDE levels in Herring Gull eggs over a period of three years after the ban on use of DDT in Norway in 1970, but found marked variation in levels between localities.

Voluntary restrictions on the use of PCBs in situations where environmental contamination could result came into effect around 1971, and levels of PCBs in North Atlantic seawater were found to decline almost immediately (Harvey *et al.*, 1974). Levels in marine organisms remained high. Levels in Great Skuas and their eggs collected in Scotland and Iceland between 1969 and 1983 suggest that reduced use of DDT and dieldrin has resulted in reductions in these pollutants in the marine food chain. DDE and dieldrin levels in Great Skua eggs show a decline, as do levels in adult muscle tissues (Table 6.6). However, results for PCBs are less clearcut. Analyses of eggs suggest a decline in PCB levels between 1974 and 1983 while analyses of adult muscle tissues suggest that PCB levels have continued to increase from 1971 to 1980 (Table 6.6).

Golden Eagles on Rhum, Inner Hebrides, are also top predators, feeding on Fulmars, Manx Shearwaters and gulls to a considerable extent. They also show signs of decreasing levels of DDE, but increases in PCBs. DDE levels (ppm wet weight) in eggs from Rhum Golden Eagles were 3.06 in 5 eggs in the 1960s, 2.3 and 2.7 in 1975, but only 1.4, 1.7 and 2.1 in eggs collected in 1984 and 1985. In contrast, PCB levels (ppm in lipid) were 3.0 in 1971, 12.4 and 13.4 in 1975 and 239, 248 and 294 ppm in 1984 and 1985. These data suggest that, despite a knowledge of the toxic nature of PCBs and some restrictions on use and evidence of local decreases in PCB levels in some areas, levels of PCBs in marine food chains have continued to increase on the eastern side of the Atlantic.

6.5 Heavy metals

6.5.1 *Definition and global distributions*

Heavy metals are often defined as metals with atomic number greater than that of iron, or as toxic metals and metalloids. The metals iron, copper, zinc, cobalt, manganese, molybdenum, selenium, chromium, nickel, vanadium, arsenic and tin are biologically essential, particularly as constituents of enzymes. In excess quantities they are toxic. Non-essential metals, including silver, aluminium, cadmium, mercury,

Table 6.6 Concentrations of organochlorines in the eggs and muscle of Great Skuas from Iceland and Scotland.

Tissue	Locality	Year	Number sampled	Organochlorine levels (ppm wet weight) PCBs	DDE	Dieldrin
Eggs	Scotland	1969	1	25.0	nd	nd
	Iceland	1973	13	27.0	5.9	0.20
	Scotland	1974	—	36.0	3.6	0.05
	Scotland	1976	12	17.6	1.7	0.08
	Scotland	1983	6	6.1	0.6	0.04
Muscle	Scotland	1971	5	15.9	2.8	nd
	Iceland	1973	10	16.0	4.4	0.12
	Scotland	1975	6	22.0	3.8	nd
	Scotland	1980	20	32.6	1.9	0.04

Note: nd = not determined

Table 6.7 Heavy metal concentrations in oceanic seawater, modern and ancient snow and ice (from Bryan, 1984; Wolff and Peel, 1985)

	Concentration (picograms per gram)		
	Lead	Mercury	Cadmium
Source			
Modern oceanic seawater	1–15	11–33	15–118
Ancient Greenland ice	1.4	< 10	—
Modern Arctic snow	200	< 10	5
Ancient Antarctic ice	1.2	< 2	—
Modern Antarctic snow	5	< 10	0.26

lead, strontium, and titanium are also toxic, because they may bind to, and thus block, sites where essential metals are required. Metals are natural components of seawater, and remain in a state of dynamic equilibrium since the amounts taken up by animals, plants or sediments are approximately equalled by the amounts entering the seas and oceans by such natural processes as erosion, volcanic activity or leaching from the atmosphere. However, because the natural levels are extremely low, the possibilities for pollution from anthropogenic (man-made) sources of metals are considerable. The main concern is the risk of human health hazards from consumption of contaminated marine foods. Since oceans are the sink for many wastes that include metals, seabirds as top predators may provide a useful monitor of the degree of contamination of marine ecosystems on a broad scale. This is particularly so for metals that can be measured in the plumage of seabirds, since this allows retrospective examination of levels at times prior to anthropogenic contamination by analysis of specimens in museum collections.

Lead, mercury and cadmium are the metals generally considered to present the greatest hazards (Bryan, 1984). Analytical techniques are still improving, but have only in the last six or seven years become adequate for accurate measurements of heavy metal concentrations in seawater, ice or snow samples (Table 6.7). This has made it almost impossible to detect long-term trends in metal concentrations. Analyses of cores of ice from the Antarctic or Greenland could, in theory, be used to measure global contamination by air-borne routes, but attempts to detect changes in levels have been hampered by the exceedingly low levels of metals in such samples and by sample contamination between collection and analysis. The most recent review of this subject concluded that there are no reliable data for heavy metals in ancient Antarctic ice, and that data from Greenland may

Table 6.8 Lead concentrations (ppm dry weight) in seabirds (from Hutton, 1981 and NERC, 1983)

Species	Tissue	Mean lead level (ppm dry weight)
Herring Gull	Bone	37.7
Oystercatcher	Bone	14.2
Great Skua	Bone	4.5
Cormorant	Eggs	< 0.2
Auks	Fat	< 1
Auks	Eggs	< 0.1

also be inaccurate due to contamination problems (Wolff and Peel, 1985). At present, seabird samples provide an alternative method of assessing changes in metal levels with time, since bioaccumulation and biomagnification result in metal concentrations in seabird tissues that are three to six orders of magnitude higher than found in water samples. For example, levels of mercury in Great Skua livers or feathers are generally in the range 1–10 ppm (μg g^{-1}) while levels in seawater are in the range 0.00001–0.00003 ppm.

Estimates of the global inputs of metals into the oceans suggest that anthropogenic input is 18 times greater than natural input for lead, 9 times greater than natural input for cadmium, and somewhere in between these two values for mercury (Bryan, 1984). It seems likely that these three metals provide the greatest hazard to marine ecosystems and, through these, to man.

6.5.2 Occurrence and distribution of metals in seabirds

Lead levels in seabirds have received little attention. Concentrations of lead in eggs and tissues of seabirds tend to be low. Bone lead concentrations tend to be much higher than levels in soft tissues. These were higher in Herring Gulls than in Great Skuas or Oyster-catchers (Table 6.8), and this was attributed to the gulls feeding on refuse tips where lead contamination is prevalent (Hutton 1981).

Mercury and cadmium levels have been the subject of several studies, and have often been examined in conjunction with copper, selenium and zinc, essential elements with which they tend to in-teract. Mercury and cadmium levels tend to be highest in pelagic seabirds and show differences within a species between populations in different localities (Tables 6.9, 6.10). High levels of metals may be associated with the type of diet; species feeding on squid seem

Table 6.9 Cadmium levels in the kidney (ppm wet weight), mercury levels in the liver (ppm wet weight) and principal diet of some seabird populations at Gough Island, South Atlantic. Species ranked according to mean cadmium levels.

Species	Number measured	Principal diet	Cadmium in kidney		Mercury in liver	
			Range	Mean	Range	Mean
Wandering Albatross	2	Squid	127–148	137.5	266–271	268
Great Shearwater	13	Squid	38–161	80.5	0.8–6.5	2.0
Sooty Albatross	8	Squid	58–92	76.4	80–227	141
Rockhopper Penguin	12	Crustaceans	32–112	72.1	1.0–3.7	2.3
Atlantic Petrel	13	Squid	42–102	63.5	14–53	26
Soft-plumaged Petrel	18	Squid	32–90	50.3	3.6–103	22
Kerguelen Petrel	13	Fish/squid	22–68	46.1	1.9–6.8	4.7
Little Shearwater	13	Squid	23–71	45.5	0.6–1.6	1.2
Broad-billed Prion	31	Copepods	19–72	34.7	0.1–1.1	0.4
Common Diving Petrel	17	Crustaceans	17–74	33.5	0.2–1.5	0.5
Yellow-nosed Albatross	9	Fish/squid	15–46	25.0	4.8–21	9.4

to have higher levels than species feeding on crustaceans. Differences between localities may be due to local effects of pollution or to differences in diets between areas. Unfortunately we do not yet know enough about natural patterns of metal accumulation to distinguish polluted levels from natural high levels.

Cadmium concentrations are always highest in the kidney, where it is thought that a specific metal-binding protein (metallothionein) generally renders the metal harmless. Mercury is more evenly distributed, since much of the body burden is in the form of methyl-mercury, which is lipid-soluble. Lead is mainly incorporated into the bones, although birds suffering from lead pollution have high levels in the kidney and the blood.

6.5.3 *Toxic effects of metals*

Mercury and cadmium are highly toxic to many marine invertebrates and to the eggs, larvae and adult stages of fish (Bryan, 1984). Both metals can cause damage to the vertebrate kidney, and mercury can also cause the infertility of eggs, is a teratogen (causes developmental mutations), and can disrupt the central nervous system, causing effects ranging from abnormal behaviour to convulsions and death (Bunyan and Stanley, 1982). In seabirds, there is little evidence to suggest that mercury or cadmium levels are high enough to cause harm, although Nicholson and Osborn (1983) reported that examination of the kidneys of Fulmars, Manx Shearwaters and Puffins by electron microscopy indicated pathological features similar to those induced in captive Starlings treated with cadmium or mercury in their diet, and not found in birds given a diet free from heavy metals. It is not clear whether this tissue damage presents a significant cost to the individual or whether seabirds are capable of continuously replacing kidney cells without causing stress to the bird.

Table 6.10 Mercury levels (ppm dry weight) in eggs of Gannets (from Nelson, 1978)

Locality	Number measured	Mercury level (ppm dry weight) mean and standard error	
Scar Rocks, N Irish Sea	18	10.5	0.71
Ailsa Craig, SW Scotland	29	4.5	0.36
Bass Rock, SE Scotland	18	2.6	0.17
Little Skellig, SW Ireland	7	3.2	0.35
Nordmjele, N Norway	10	2.9	—

6.5.4 *Feathers as a monitor of metals*

Although feathers are predominantly composed of keratin, many elements enter feathers from the blood during feather growth. This is particularly the case with metals, many of which bind strongly to sulphur atoms in sulphur amino acids that are especially abundant in keratins. Metal levels in feathers are determined by levels in the blood at the time of feather growth, although surface contamination can cause changes after feather growth (Goede and de Bruin, 1984). Feather washing can remove most surface contamination while having little effect on levels of heavy metals bound to feather proteins during growth. Thus feathers may be used to measure metal levels in seabirds, providing the pattern of feather replacement is taken into account (Furness *et al.*, 1986). Feathers have been used particularly to measure mercury levels and, since mercury bonding to feather proteins is strong, stored feathers (e.g. from museum skins) may be used to determine historical levels (Appelquist *et al.*, 1984). Measurement of levels in feathers from museum specimens has shown that mercury levels in Guillemots and Black Guillemots have increased over the last 150 years in the Baltic (Figure 6.5) and probably also to a smaller extent in the North Atlantic (Appelquist *et al.*, 1985). One problem with using museum specimens to monitor mercury levels is that mercuric chloride was often used as a preservative by taxidermists, so that museum skins may be heavily contaminated by inorganic mercury. Since the mercury in birds is almost all organic, methyl-mercury, it is possible to discriminate between these two categories by analysing levels of methyl-mercury in the feathers. Feather analysis has great potential in studies of mercury in marine ecosystems, since feathers are readily available without having to kill birds and museums hold long time-series of many species.

6.6 Radionuclides

We know very little about the uptake and transfer of radioisotopes by marine organisms and their transfer to seabirds at the top of the food chain. Since the Irish Sea has become the most radioactive sea in the world, there is interest in the possible effects of accumulation of isotopes in seabirds from this area. In recent years there have been considerable decreases in numbers of Black-headed Gulls nesting

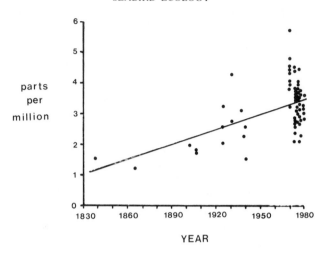

YEAR

Figure 6.5 Levels of mercury (ppm dry weight) in the sixth primary of Black Guille-mots and Guillemots from the Baltic Sea collected between 1830 and 1980. (From Appelquist *et al.*, 1985.)

on the Cumbrian coast, and it has been suggested that these could be related to increased uptake of radionuclides. There is as yet no evidence to support this suggestion, but it clearly merits attention since this form of pollution is one to which marine ecosystems have not previously been exposed and so it is difficult to predict what accumulations and effects might occur.

6.7 Monitoring fish stocks

At present, it is uncertain whether seabirds may be used to provide information on the status of fish stocks. Some studies suggest that this may be possible. For example, analysis of the anchovies con-sumed by Jackass Penguins may allow an estimate of the rate of recruitment of juvenile fish to the commercially exploitable fraction of the population, since juvenile anchovies are eaten by penguins earlier in the year than they appear in commercial catches. Anchovies occur in closed (concentrated) or in open (dispersed) shoals. Only closed shoals are commercially exploitable. While Jackass Penguins prey mainly on open shoals, Cape Cormorants prey primarily on closed shoals. Comparison of the diets of these two species may provide information on both prey population dynamics and behaviour (Hockey *et al.*, 1983).

Numbers of terns breeding in the Clyde sea area, Western Scotland, correlate with the catch per unit effort of herring two years later (Monaghan and Zonfrillo, 1986), implying that counts of terns would provide an index of herring recruitment into the commercial fishery as two-year-olds. In this case, presumably tern breeding numbers must correlate with the abundance of O-group herring (less than one year old) when terns start to nest.

Hislop and Harris (1985) demonstrated that the average annual proportions of herring and sprats in the food loads of Puffins brought to chicks at the Isle of May, Firth of Forth, correlated with estimates for the same year of larval herring abundance and sprat total biomass in the North Sea obtained by fishery research surveys. While they concluded that dietary data from seabirds might not be a reliable enough method for estimating strengths of recruiting year classes of sprats and herring, they suggested that seabird diets might be of value as broad indicators of trends in fish abundance.

Since none of these studies set out to determine the potential use of seabirds as monitors of fish stocks, it seems likely that a study directed to this objective would show that certain seabirds could be used to provide an inexpensive 'early warning' system indicating recruitment failure or stock depletion.

6.8 Monitoring seabird numbers

The aspects of seabird population biology most sensitive to environmental changes are probably diet, adult activity budgets and breeding success. However, since considerable sustained effort is required in order to monitor these, most monitoring programmes have been based on counts of seabird numbers. This is most appropriate where seabird conservation is the issue, since changes in diets, activity budgets or even breeding success need not necessarily result in changes in breeding numbers. A major drawback with monitoring numbers is that population counts alone provide no information on the cause of any change detected. Presumably the rationale for this is that serious changes will be detected by monitoring numbers, and once detected will lead to concentrated efforts to determine the cause of change.

Despite the considerable efforts and resources put into seabird monitoring in recent years, the objectives of monitoring studies have rarely been stated explicitly, and rather little attention has been paid to the sensitivity of such studies. Some could detect only catastrophic changes, while the design of some others may lead to spurious trends being measured.

In some cases, populations have been monitored by counts of breeding numbers in every colony at well-spaced time intervals (e.g. decadal counts of Kittiwakes and Fulmars in Britain and Ireland: Coulson, 1983; Fisher, 1966). The main limitations of this are that only long-term trends can be detected, and it is difficult to achieve thorough coverage. The biggest advantage is that no sampling procedure is required so that changes detected are representative of the population.

In order to allow monitoring of seabirds that breed in large numbers at many localities, sample monitoring plots have been established where defined parts of cliffs are monitored in detail, usually on an annual basis. A serious problem with this method is that selection of sample plots may determine the results of monitoring. For example, plots at the centre of colonies are unlikely to show increases or decreases in numbers to the same extent as plots at the edge. Because cliffs and slopes inhabited by burrow-nesting seabirds are often fragile and erode over the years, colony boundaries may move so that numbers in monitoring plots may decrease or increase as a result of relocation of nest sites, even when total numbers are not changing.

Annual monitoring of diets and breeding success of seabirds might be much more informative than monitoring of numbers, but this has received little consideration to date. There is a clear need to reappraise the concept of seabird monitoring so that future programmes will be able to detect changes in seabird numbers and biology and relate these to causal environmental factors.

CHAPTER SEVEN

SEABIRDS AS PESTS

Animals come to be considered pests when they interfere with the way in which we utilize our environment, for example by competing with us for resources, damaging our constructions or spreading disease. The more successful they are in so doing, the more seriously we consider them as pests. For obvious reasons, the greater the local population size of a particular pest species, the less it tends to be tolerated.

Many seabird populations have expanded considerably during the present century, and the factors affecting the regulation of seabird populations have been discussed in detail in Chapter 4. In Europe and North America, the increase in several gull species has been especially marked (e.g. Spaans 1971, Drury 1973, Cramp *et al.* 1974). The Herring Gull in Britain has increased its numbers at an average annual rate of 13% between at least 1940 and 1970 (Chabrzyk and Coulson, 1976), which means in effect that its population doubled every six years (Figure 7.1). As with many other seabirds, this population expansion was partly a recovery from persecution in response to legislation introduced to protect seabirds from the excessive exploitation for sport and the plumage trade in the nineteenth century (see 8.1), and a consequence of changing food supplies. However, in addition, gulls have benefited greatly from their abilities to invade new habitats and exploit new food sources (Conover, 1983; Monaghan, 1983), and it is these aspects of their behaviour which had the result that they now cause the majority of seabird pest problems.

7.1 Conflict with fisheries and agriculture

Competition between nations for marine fish resources in many areas of the world has led to concern that top predators such as

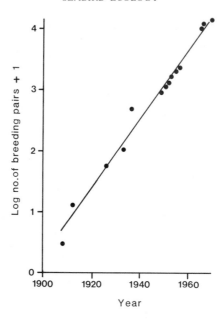

Figure 7.1 Changes in the number of Herring Gulls breeding on the Isle of May, Scotland. The average annual rate of increase during this period was 13% per annum. (Data from Chabrzyk and Coulson, 1976).

seals and seabirds eat large quantities of fish which would otherwise provide an economic profit for fishermen. To assess the extent of this competition we need realistic estimates of the nature and amount of food seabirds consume and how this relates to fisheries yields. Such studies are hampered by a fundamental lack of detailed knowledge of the size and distribution of relevant fish stocks in relation to seabird distribution. These seabird–fisheries interactions have been dealt with in detail in Chapter 5.

Another area of potential conflict between seabirds and fisheries relates to the increasing development of fish farming, especially of salmonids. Seabirds which occasionally feed in fresh water, such as cormorants, will prey on young fish in rearing tanks and cages at the farm itself; in addition, they often inflict sufficient damage on surviving fish to render them unmarketable. Gulls, though often unable to reach the fish directly, will feed extensively on the expensive fish food. Furthermore, while gulls, terns and auks eat salmon smolts when they are abundant, the distribution pattern of wild smolts appears to be such that they do not usually represent a profitable prey for seabirds. However, the unique conditions pertaining at a hatchery release site, where large numbers of naive smolts are released into a river mouth, may make this food source more profitable. As fish farming expands in economic importance, this potential area of conflict between seabirds and man may increase.

With regard to agriculture, there is comparatively little conflict between seabirds and man. The main problem area relates to the extent to which inland feeding gulls are involved in the transmission of pathogens to domestic animals (see 7.4). Large seabirds such as Great Black-backed Gulls and Great Skuas are accused of stealing new-born lambs in some areas, though available evidence suggests that this is comparatively rare and, where it does occur, involves mainly weakened lambs unlikely to survive in any event. The aggressive behaviour of the Arctic Skua in defence of its nest and young has led to crofters on Fair Isle, Shetland complaining of being attacked themselves—one of the few cases of direct conflict between seabirds and man (see 7.3.1)! However, the relationship between seabirds and farmers is not always so acrimonious, since gulls may sometimes prey on invertebrate agricultural pests. It was for such an invaluable service rendered to the Mormon community in the early days of settlement in the United States that a monument to the California Gull was erected in Salt Lake City, and the species is afforded special protection in the area.

7.2 Detrimental effects on other species

The population expansion of many of the large gulls has resulted in their having potentially damaging effects on the vegetation of their breeding colonies. They may also affect the breeding success of other seabirds by competing with them for nesting space, stealing their food or preying on their eggs and young.

The effect of large gull colonies on the soil and vegetation of their nesting sites has been studied in several localities. In general, the typical maritime vegetation is replaced by coarse grass and weedy annual species. This results from a combination of factors, including the deposition of guano which causes major changes in soil conditions, physical damage during grass pulling displays and nest-building, trampling, and the transport of weed seeds by the birds into the nesting area (Sobey, 1976; Hogg and Morton, 1983). However, while these changes could be said to render the habitats less attractive to human visitors, this is not really a serious problem. Other seabirds, particularly burrowing species such as Puffins, can also have detrimental effects on their breeding habitat, in this case by greatly increasing soil erosion.

Direct competition for nesting space does occur between species with similar nesting requirements, especially where there is a difference in the timing of colony occupation. For example, former breeding sites of the rare Audouin's Gull in the Mediterranean have been taken over by the expanding population of Herring Gulls which set up territories 2–3 weeks before the Audouin's Gulls arrive (Bradley and Monaghan, in press). Gulls have also taken over former breeding sites of terns in many areas, and tern breeding success can be adversely affected by gull predation (Thomas, 1972; Becker, 1984). However, the extent to which gulls actually drive out terns is unclear, and in some areas both have increased in parallel, while in others terns have declined in line with national trends but in the absence of gull predation (Becker and Erdelen, 1985; Monaghan and Zonfrillo, 1986).

Studies of the effect of kleptoparasitism by gulls on the breeding success of the host species such as puffins and terns suggest that this is not a major factor (Veen, 1977; Pierotti, 1983) and in the absence of undue disturbance, most seabirds can protect their eggs and young to some extent from gull predation. Indeed they may receive protection from more serious predators by nesting in association with gulls, as is the case for Sandwich Terns nesting with Black-headed Gulls (Veen, 1977).

7.3 Gulls and the urban environment

Gulls, unlike most other seabird species, have generally benefited from their association with human populations. Their lack of extreme specialization to the marine environment has allowed them to exploit resources provided by man for feeding, breeding and nesting. Species such as Herring and Great Black-backed Gulls, hitherto restricted to coastal environments, have increased their utilization of urban areas during the present century. This has brought them well inland but, as their numbers have expanded, has also brought them into increasing conflict with human populations.

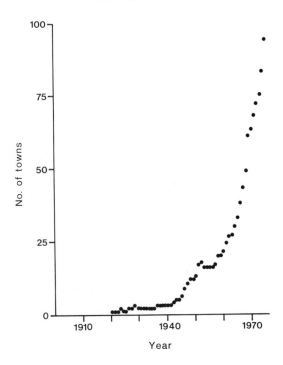

Figure 7.2 Changes in the number of towns in Britain with breeding colonies of Herring Gulls on buildings. The increase is exponential. (Data from Monaghan and Coulson, 1977.)

7.3.1 *Urban nesting*

Though first documented early this century, gulls breeding on inhabited buildings in towns and cities were rare before 1940. The habit has since increased considerably (Figure 7.2). In Britain for example, the number of rooftop nesting Herring Gulls increased from around 1200 pairs in 1969 to more than 3000 pairs in 1976, an average annual increase of 17% (Monaghan and Coulson, 1977). Herring Gulls are attracted to towns by the available food sources; a shortage of suitable breeding sites due to saturation or near-saturation of nearby breeding colonies causes them to take up nesting on the town buildings. This is probably not such a radical departure from their normal breeding sites as it may initially appear since, to a Herring Gull at least, the outline of the buildings in

Figure 7.3 A dense flock of Herring Gulls feeding on newly-tipped refuse.

a town may resemble the irregular rocky outcrops characteristic of their traditional coastal sites. In addition to the proximity to food sources, the spacing imposed by the structure of the buildings improves their breeding success (see 2.3.3).

Although only a comparatively small number of birds are involved, gulls breeding in towns do cause considerable problems. Noise, damage to the fabric of buildings, fouling of people, pavements and buildings, and their dive bombing behaviour in defence of their nest and young all give cause for complaint in towns; there is also some concern that they may be involved in the spread of disease (see 7.4).

7.3.2 *Gulls and refuse tips*

The changeover from the widespread, haphazard collection of refuse throughout urban areas, to its organized collection and tipping at central locations, has provided gulls with an abundant food source throughout the year. The construction of large water-storage reservoirs in some areas has facilitated their exploitation of inland tips, by providing suitable roosting sites sufficiently close to the feeding area. The collection of refuse in Britain was instituted by the Public Health Act of 1875, and by 1922 complaints concerning gulls feeding at tips were already being made. Herring, Great Black-backed, Lesser Black-backed and Black-headed Gulls all now make extensive use of tips, as indeed do most of the large gulls in other countries where such sites are available. However, there is some regional variation in the extent to which the same species utilizes refuse tips in different areas. For example, the Black-headed Gull commonly feeds at tips in north-east England but rarely does so in west and central Scotland where it concentrates more on coastal, estuarine and agricultural areas.

More than 15 million tonnes of household refuse are currently produced annually in Britain, 18% of which is vegetable and putrescible matter. An increasing proportion is disposed of by incineration, but by far the most common disposal method is still landfill tipping, as in most other countries. Refuse at these sites is deposited by collection lorries, and then compacted by bulldozers into shallow layers and covered with earth. Different tipping practices may favour different gull species. In New Jersey, USA, for example, the increased bull-dozing and covering of refuse is thought to favour the more agile Laughing Gulls, which dip for food in flight and can do this while the refuse is being moved; their increased use of tips in this area has been attributed to this change in tipping practice (Burger, 1981). Similarly, the smaller and more agile female Herring Gulls make more use of this feeding method at tips than do the males (Greig et al., 1985).

Refuse tips are used mostly by gulls during the winter period, and feeding at tips tends to be highly competitive, with dense flocks collecting on areas of newly-tipped refuse (Figure 7.3). Larger gulls tend to dominate smaller individuals and occupy the best feeding positions; this is true both within and between species (Greig et al., in press). However, female Herring Gulls feed more at refuse tips than males, especially during winter, and this affects the extent to which they carry pathogens (see 7.4.1).

The availability of refuse has affected both the winter distribution of gulls and the extent to which populations have been able to ex-pand. The problems they cause at these sites mostly relate to noise and the dispersion of inedible scraps over the surrounding areas.

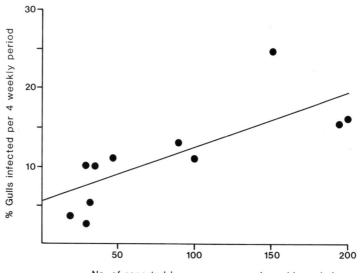

Figure 7.4 The proportion of Herring Gulls carrying Salmonellae in relation to the number of human cases in the same area at the same time. There is a significant correlation between the two ($r = 0.78$, d.f. 9, $P < 0.01$). Data from Monaghan *et al* . (1985).

Where a tip is adjacent to pastureland, there is some concern over the possible spread of disease to domestic animals; the contamination of potable water supplies by roosting birds may also occur (see 7.4).

7.3.3 *Bird strikes*

A 'bird strike' refers to a collision between an aircraft and a bird or bird flock. Since birds rarely fly above 500 metres such collisions are most likely to occur during aircraft landing or takeoff, or during low flying operations. Damage to aircraft is most severe when birds are sucked into jet engines, as illustrated by an incident in December 1985 at Dublin Airport; a single gull which was sucked into the engine of a passenger jet before take-off resulted in almost instant, complete power failure, and more than one million pounds of damage to the aircraft. Where such strikes occur after take-off, human lives have been lost as a result.

In Britain, one bird strike occurs for every 1500 flights by civil aircraft, and the incidence is higher for lower flying military planes.

Gulls are involved in more than 40% of all bird strikes (Rochard and Horton 1980). Gulls, like other birds such as lapwings, are attracted to airfields by high invertebrate densities in the short grass areas, and the wide green spaces offer good loafing areas. Growing the grass longer helps to alleviate this problem (Brough and Bridgeman 1980). They may also fly to and fro across airfields, commuting between feeding sites such as refuse tips and roosting sites. Deterrents such as distress calls (see 7.5) are used to scare birds away, but appropriate siting of refuse tips in relation to airports is a very important aspect of gull control in this context.

7.4 Disease transmission

7.4.1 *Salmonella carriage by gulls*

In recent years, there has been increasing concern over the role of gulls in the dissemination of pathogens such as the Salmonellae. These organisms commonly cause enteric disease in humans, usually referred to as 'food poisoning'. Gulls can ingest these pathogens at feeding sites such as refuse tips and sewage outfalls, and then void them in their faeces. Thus, where gulls roost on pastureland (by day), or water storage reservoirs (by night) or nest on inhabited buildings, they may be in a position to transmit these pathogens to man or domestic animals.

Numerous *Salmonella* serotypes have been isolated from the faeces of gulls, and in general these mirror those prevalent in human populations. A recent study in the Clyde area of Scotland found that around 10% of the Herring Gulls feeding at refuse tips carried *Salmonella*, though carriage rates were often considerably higher in females during winter due to their increased dependence on refuse tips (Monaghan *et al.*, 1985). Geographical variations in carriage rates can also be linked to the feeding ecology of the birds and in general carriage rates in gulls are higher in areas of high human population density (Girdwood *et al.*, 1985).

There is a positive relationship between the incidence of *Salmonella* carriage in gulls and the incidence of infection humans in the same area at the same time (Figure 7.4). However, this relationship does not arise because gulls infect humans; rather, the reverse is the case. Gulls carry only small numbers of Salmonellae

F

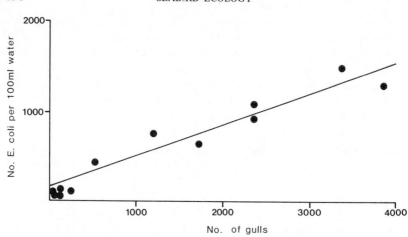

Figure 7.5 The relationship between the number of Herring Gulls roosting on a Glasgow reservoir and the degree of contamination of the water with faecal bacteria over one year. There is a significant correlation between the two ($r = 0.96$, d.f. 10, P I 0.001.) Data from Benton *et al.* (1983).

for short periods after ingestion and are not actively infected. They come in contact with these pathogens as a result of the way we dispose of our sewage and refuse, and thereby reflect the level of environmental contamination. However, since they can travel several hundred kilometres in only a few days, they may be involved in the transport of pathogens between different areas. Individual gulls do not generally carry sufficient *Salmonella* to infect healthy humans or domestic animals. Nonetheless, when large numbers of gulls are present, they can contaminate water to an appreciable extent. An infection may result if water from a pond used by gulls is drunk by livestock rendered susceptible due to stress or other causes (Reilly *et al.*, 1981).

The presence of large winter gull roosts on water storage reservoirs is a problem for many water authorities. Such roosting flocks can be very difficult to deter, though some success has been achieved by stringing spaced wires across the water, often a prohibitively expensive undertaking at large reservoirs (Amling, 1980). The problem of roosting gulls on water storage reservoirs in the city of Glasgow was successfully tackled by Benton *et al.* (1983). The water storage reservoirs for this city had been used as a nocturnal roost by gulls over at least 30 years. These reservoirs now supply over 900 000 consumers and the quality of the incoming water was initially so high that the only purification facilities required were marginal chlorination and pH correction. However, the bacteriological quality of the water declined steadily over the years as the number of gulls using the reservoir increased; in addition, the contamination was highest during winter, correlating with the presence of the roosting gulls (Figure 7.5). Deterring the roosting gulls by controlled use of distress calls (see 7.5) resulted in a marked improvement in water quality, and thereby saved the expenditure of tens of millions of pounds in upgrading the treatment plant.

7.4.2 Other diseases

In addition to the Salmonellae, gulls can also carry other pathogenic organisms, notably *Campylobacter*, which also cause enteric disease in man. They have also been implicated in the spread of the beef tapeworm by transporting eggs from sewage to pastureland, but there is little direct evidence that this occurs.

Although 'ornithosis' (or 'psittacosis' as it was originally termed), is traditionally associated with parrots, it is also carried by many other birds including seabirds (Miles and Shrivastav, 1951). Humans can become infected with this potentially serious respiratory disease by handling infected birds or inhaling their dried droppings. For example, an epidemic of this disease occurred in the Faeroes in 1938, largely amongst women who were involved in the splitting and salting of Fulmars trapped for food; of 165 cases, over 19% were fatal (Fisher, 1952). However, ornithosis is usually only a mild infection and responds well to antimicrobial drugs.

A large number of viruses have also been isolated from seabird ticks. These 'arboviruses' may occasionally cause disease in seabirds, but for the most part are not known to cause serious disease in man (Nuttal, 1984).

7.5 Control of seabird pests

As with other animals, in controlling gulls and the problems they create, it is necessary to take account of their behaviour and ecology (Monaghan, 1984). Attempts to control breeding gulls by lowering their reproductive output is likely to have little success, especially in the short term, since they are long-lived with a low annual reproductive rate. In addition, their deferred maturity means that, at any one time, the young birds which will recruit into colonies over the next five years are already in existence. Culling of breeding adults, if it is to be done at all, must not be done in a way that maintains the birds at low density, since this will remove the social constraints which may limit recruitment (see 2.3.3); other density-dependent effects may also come into play (see 2.7).

Deterrents can be used successfully in some situations although gulls readily habituate to loud noises. The broadcasting of taped distress calls (the harsh scream given by birds when captured by a predator) can be used successfully in some situations (see 7.4.1). Provided the call is not played on a random basis, habituation is less of a problem than with other deterrents; however, where the birds are willing to take risks to defend or obtain a resource, such as a nest site or much needed food source, these calls are not very effective.

CHAPTER EIGHT

SEABIRD CONSERVATION REQUIREMENTS

The small clutch size, low breeding success, higher juvenile than adult mortality and prolonged period of immaturity of seabirds makes them particularly vulnerable to any factors that increase the mortality rate of adults. Since these characteristics mean that there is limited scope for density-dependent compensations in natality or recruitment, an increase in adult mortality rate is likely to lead to population decline. Although age at first breeding may be reduced when population size falls (see 2.6), and breeding success may increase, these can only change by small amounts, so could only balance very slight increases in adult mortality rates. By contrast, a reduction in natality may have a very small effect on population size over a long period because the annual survival rates of adult seabirds are often extremely high. If all the 50 000 or so pairs of Fulmars at St Kilda, in the Outer Hebrides, failed to fledge another chick for the next 20 years and no immigration took place from other colonies, the number of pairs of Fulmars at St Kilda could remain the same for the first ten years of reproductive failure because chicks fledged in earlier years would continue to recruit for at least ten years (probably for nearer 20). Then numbers would decline by only about 3% per annum (the adult mortality rate), so that after 20 years of total reproductive failure and assuming that St Kilda was a closed population, there would still be about 37 000 pairs of Fulmars remaining. Given that, in reality, numbers fluctuate a little from year to year for a variety of reasons, this reduction might be quite difficult to detect, and since we know that immigration and emigration are also important processes, we can see that massive breeding failures at some colonies may actually have very little effect on the size of the breeding population. Of course, seabirds with lower adult survival rates and higher fecundity (such as cormorants, terns, Peruvian guano seabirds) will be somewhat more sensitive to reductions in breeding performance. Nevertheless, it is threats to adult seabirds that present the greatest conservation dangers (Table 8.1). Many seabirds are subject to several threats acting together, and in such instances it is difficult to decide which is the cause of a population decline. For example, the 90% reduction in numbers of Jackass Penguins in southern Africa has been attributed by different authors to human exploitation for eggs, oiling at sea, overfishing of pelagic fish stocks, disturbance of breeding colonies by guano workers and tourists, and destruction of nesting habitat. In addition, some birds are caught and drowned in fishing nets, and some colonies have been affected by introduced alien or natural predators, but these effects have been considered small in comparison to the other factors.

Even when only a single factor is identified as a threat to a declining population of seabirds, it is usually a case of interpreting a correlation as a causal relationship. This is a very risky procedure, and has led to numerous assertions, based on anecdotal evidence, that come to be accepted through repetition as if established fact. Some of these are considered in the following sections where we outline specific threats to seabirds.

8.1 Exploitation of seabirds by man

Seabirds are exploited by man in many parts of the world, primarily for food, but also for their oils, feathers, and as bait for fishing. In many regions the level of harvesting has decreased, but in remoter areas exploitation has increased as more sophisticated transport and harvesting methods have become available, and seabird products have become an exportable resource as well as a subsistence staple. For example, gulls' eggs traditionally harvested for local consumption by the sparse communities in arctic Norway are now for sale in Oslo at much higher prices, and amongst other things, served as a delicacy to passengers on Scandinavian aircraft.

In many cases the extent of exploitation is poorly documented, and as a consequence we can only speculate as to the effect on seabird populations. In some of the more important and extensive cases, documentation is much better (Table 8.2).

The now extinct Great Auk, as well as being flightless, had a very restricted breeding distribution, and tended to occur in relatively small numbers compared to other auks, making it particularly vulnerable to exploitation. Huge, though largely undocumented, numbers were killed by fishermen who provisioned long voyages with salted Great Auks between AD 1500 and the early 19th century, and it has generally been accepted since the extinction of the species in 1844 that this exploitation caused its demise. However, this apparently clearcut conclusion has recently been challenged. There is good evidence that a climatically severe period preceded and coincided with the period of exploitation by man, and there is some evidence to suggest that Great Auk distribution had contracted before extensive exploitation began. In the absence of data on rates of exploitation and on Great Auk population dynamics we cannot rule out the possibility that climatic factors were of some importance, and possibly the major factor causing the extinction of the species (Bengtson, 1984). However, while climatic change could have been a contributory factor, the mass slaughter of adult Great Auks by man is likely to be the main reason for the extinction, since long-lived seabirds can tolerate prolonged periods of reproductive failure, and could presumably move south with their prey populations when climate deteriorated. Our knowledge of population dynamics theory tells us that greatly increased adult mortality rates could not be sustained by such a long-lived species with naturally low fecundity.

It is difficult to determine quantitatively what level of exploitation or destruction seabird populations can tolerate. Feare (1984) suggested that cropping at levels above natural mortality levels would clearly be detrimental, but cropping at levels below or equal to natural mortality would probably be acceptable. The rationale for this is that for populations limited in a density-dependent way, natural mortality would be reduced by an amount equalling mortality due to man. Thus natural mortality and exploitation would be substitutive rather than additive. While this may be the case, as we have seen in Chapter 4 we do not know for certain that seabird populations are regulated by density-dependent mortality, so this rule of thumb is somewhat speculative, and possibly rather optimistic. To take a particular case, Feare (1976a,b) showed that in undisturbed colonies, 42% of Sooty Tern eggs laid failed to produce a fledgling, suggesting that up to 42% of Sooty Tern eggs could be harvested by man without harm to the population, but in fact Feare suggested that in this case cropping levels of no more than 20% should be allowed. We

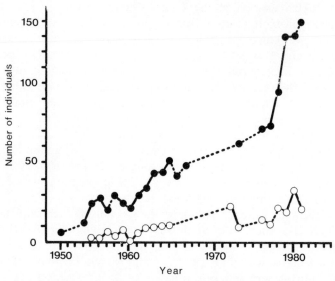

Figure 8.1 Number of Short-tailed Albatross nesting on Torishima, Japan.

need a better understanding of seabird population dynamics before safe exploitation levels can be determined with confidence.

The Short-tailed Albatross once nested in huge numbers on about a dozen remote, uninhabited islands south of Japan. From about 1870 to 1930 adults were harvested for their feathers, and it was thought that the species had become extinct when volcanic eruptions and cattle grazing destroyed most of the nesting habitat on the last remaining site. However, a few pairs were found breeding on Torishima in the 1940s and numbers have since increased (Figure 8.1). Clearly, exploitation of this species was excessive, and it seems likely that it survived largely because the long periods of immaturity will have meant that some young birds were dispersed at sea when cropping of adults removed the last breeders. Torishima is an active volcano, so that the only breeding site could still be lost. It is also infested with Black Rats. Albatross breeding success may be reduced by the unstable nature of the steep slopes, and the Environment Agency of Japan has initiated a programme of habitat improvement by transplanting native plants at a time of year when the albatrosses are at sea (Hasegawa, 1984).

Heavy exploitation of seabirds for feathers to be used in the millinery trade was not confined to exotic birds in the Far East or

Africa. In Britain and in the United States huge numbers of terns were shot for their feathers, or even to be mounted entire as the crowning glory of a lady's hat. Kittiwakes at colonies in northeast England were heavily persecuted, and a conservationist backlash against this slaughter of seabirds for feathers resulted in the establishment of the first bird protection organizations in Britain and the United States and the passing of legislation to protect seabirds.

In the 19th century, North Atlantic Great Skuas nested only in Iceland, the Faeroes and north Scotland. In the Faeroes they were killed as vermin to reduce their predation on seabirds such as Puffins, harvested in a controlled fashion by the islanders. The small number in north Scotland became the target of Victorian skin collectors from England and were reduced to only about 20 pairs around 1900, in spite of attempts at protection by conservationists. Collectors also shot many adults in the Faeroes and reduced that population to four pairs (one on each of four islands) in 1900. Legal protection was then given in both countries and numbers increased, in northern Britain at a rate of about 7% per year. There is little doubt that the population reductions were the direct result of shooting of adults, while the increase since 1900 in Britain has been more than just a recovery from persecution; it has led to a far larger population than was present before the period of persecution and this seems to be due to changes in feeding ecology in relation to the fishing industry (Furness and Hislop, 1981), indicating the complex interactions between factors affecting seabird population sizes.

Human exploitation caused the extinction of the Spectacled Cormorant, and almost certainly also the Great Auk, nearly drove to extinction the Cahow and Audouin's Gull as well as the Short-tailed Albatross, and nearly wiped out the Faeroese and British populations of the Great Skua. In all these cases it was exploitation of breeding adults that caused the problem. Exploitation of seabird eggs, chicks or immatures appears to have had little impact on populations. Faeroese fowlers largely exploited immature rather than breeding adult Puffins, and regulated their catches according to traditions that presumably arose in order to avoid overexploitation (Harris, 1984).

8.2 Habitat destruction

The destruction of nesting habitat is particularly serious for seabirds that breed in small numbers at only a small number of sites. Several

rare species of gadfly petrel (*Pterodroma* spp.) are threatened
by burrow destruction due to agricultural development, vegetation
clearing or introduced trampling animals such as cattle, goats and pigs.
Phosphate mining on Christmas Island in the Pacific has resulted in
the clearing of 25% of the trees in the nesting area of the endemic
Abbott's Booby, which appears unable to adapt to other nesting
sites (Nelson, 1978). Tree-nesting Frigate-birds were exterminated
from St Helena by logging, and have declined at Christmas Island
and elsewhere (Vermeer and Rankin, 1984). Terns are highly sensitive to disturbance of their breeding sites, and several species are
threatened by human development of their nesting habitats (Table 8.3).

In general, many seabird populations are affected by habitat destruction and human disturbance (Vermeer and Rankin, 1984) but
most cases are due to direct and premeditated effects of man, so
could be controlled if the requirements of conservation were perceived
to outweigh the benefits of development.

8.3 Predation by introduced alien animals

Seabirds show few behavioural adaptations to repel ground predators, and avoid predation mainly by nesting on remote islands.
Establishment of alien predators on seabird islands by accidental
or deliberate introduction has occurred frequently since man began
exploration and exploitation of remoter areas. Small seabirds are
particularly vulnerable. Imber (1975) suggested that petrels are endangered by rats only if their body weight is less than that of the
mammal, and this may apply to other seabird groups.

The severity of predation on seabirds depends on the availability
of alternative prey and the seasonality of the environment. On sub-
Antarctic Macquarie Island in the South Pacific, feral cats subsist
on the sparse population of rabbits during winter when seabirds
are absent and their numbers are correspondingly limited (Jones,
1977); by contrast, cats on Dassen Island, South Africa, prey almost
exclusively on abundant rabbits throughout the year and ignore
seabirds (Cooper, 1977). On Marion Island, in the Southern Indian
Ocean, seabirds are present throughout the year and an estimated
450 000 were killed by 2200 feral cats each year (van Aarde, 1979),
apparently causing a marked reduction in numbers of small petrels.
The reduction in seabird numbers is inferred rather than shown to
be due to predation by cats, but in this case a control island exists

adjacent to Marion Island. It is cat-free and shows no sign of a decrease in small-petrel populations there (Williams, 1978).

Cats, and rats (*Rattus rattus, R. norvegicus, R. exulans*), are generally considered to be the most serious alien predators of seabirds, and are widespread (Moors and Atkinson, 1984). These, and other predators such as mongooses, mustelids, pigs and foxes, have undoubtedly had very serious effects on populations of seabirds in many areas (reviewed by Moors and Atkinson, 1984), although the only species of seabird thought to have become extinct due to alien predation is the Guadelupe Storm Petrel.

Control or elimination of introduced alien animals is difficult. Cat numbers on Marion Island have been greatly reduced by introduction of the host-specific viral disease feline panleucopaenia, supplemented with shooting and trapping (Moors and Atkinson, 1984). Cats can be eliminated from small islands, but the success of programmes on larger islands has yet to be seen. After elimination of cats from Baker Island in the Pacific, colonies of Sooty Terns and Lesser Frigate-birds became re-established, indicating that removal of cats can have important conservation consequences (Moors and Atkinson, 1984).

Foxes have clearly affected the variety of species of seabirds on islands off Alaska, and attempts to remove fox populations have been quite successful. One method being developed is the introduction of sterile Red Foxes with which populations of Arctic Foxes are thought to be unable to coexist (Bailey, 1982).

It seems unlikely that rats can be eliminated from large seabird islands once established, so that understanding their interactions with seabirds is important. It is unwise to extrapolate from studies of rat biology in farmyards, and there is a pressing need for field studies of the ecology of rats on seabird islands.

8.4 Other factors

Oil, plastic, toxic chemicals and heavy metals may all kill seabirds, but few seabird populations appear to be at risk of elimination as a result of pollutants. The Brown Pelican is particularly sensitive to DDT and PCBs and populations in the Southern California Bight declined due to pollutant-related reproductive failure and poisoning of adults in the 1960s and early 1970s. However, reduced DDT levels have allowed this population to recover, and its breeding success now appears to be limited by anchovy abundance (Anderson and Gress, 1983).

Large numbers of Short-tailed Shearwaters and Tufted Puffins are caught in gill-nets of the Japanese salmon gill-net fishery in the North Pacific. Estimates of mortality vary depending on the method of sampling and calculation, and it is clear that mortality varies a great deal from area to area and between vessels, but it is generally believed that between 250 000 and 500 000 birds die this way each year, of which about two-thirds are shearwaters (King, 1984). It is not clear whether these are mainly adults or immatures or both. In many seabirds, immatures are more vulnerable to accidental death from tangling in nets, and losses of immatures would have less impact on the population than similar losses of adults. There are possibly over eight million pairs of Short-tailed Shearwaters breeding in Tasmania (Naarding, 1980) but it is not known whether these drowning losses have had any effect on population size, partly because shearwater colonies are difficult to census. Monofilament drift-netting for salmon and cod off west Greenland and eastern Canada also kills large numbers of seabirds; at least 100 000 guillemots are thought to drown each year (Piatt and Reddin, 1984). Again, the impact of this on populations is not clear, but numbers of guillemots in west Greenland have decreased considerably in the last 50 years and this is generally attributed to the combined effects of hunting and drowning in nets (Salomonsen, 1979). In theory, legislation should protect guillemot populations in Greenland, but efforts to reduce hunting and drowning losses have not been altogether successful (Salomonsen, 1979).

Commercial fishing may result in reductions in food available to some seabirds, for example anchovies to Peruvian guano seabirds, but this is unlikely to cause the extinction of seabird species since a severe reduction of prey stocks would result in the abandonment of the fishery until stocks recovered. More importantly, human effects on marine ecosystems are likely to alter the competitive balance between species, causing one to increase at the expense of another. Thus some tern colonies have been forced to abandon their breeding sites by population increases of larger gulls, territories of Arctic Skuas in Shetland have been taken over by Great Skuas whose numbers have been increasing due to the availability of discards from trawlers and increased sand-eel stocks; Audouin's Gulls have been losing nest sites, eggs and chicks to increasing numbers of Herring Gulls that have benefited from garbage and fishery waste in the Mediterranean.

Seabirds most vulnerable to reductions in food availability are those with limited foraging ranges (e.g. terns and penguins), specialized feeding habits, surface feeding habits, limited ability to increase time spent foraging, energetically expensive foraging techniques, and low tolerance of temporal fluctuations in food availability (Furness and Ainley, 1984).

Rarity itself leads to increased dangers. By breeding at only a small number of sites a species is more vulnerable to changes in that area. Rare species are also likely to be so because they have very specialized and restricted ways of life, making them less adaptable to change.

REFERENCES

Adams, N.J. and Brown, C.R. (1984) Metabolic rates of sub-Antarctic procellariiformes: a comparative study. *Comp. Biochem. Physiol.* **77A**, 169–173.

Ainley, D.G. and Sanger, G.A. (1979) Trophic relations of seabirds in the Northwestern Pacific Ocean and Bering Sea. In *Conservation of Marine Birds of Northern North America*, Bartonek, J.C. and Nettleship, P.N. (eds.), *US Dept. Int. Wildl. Res. Rep.* **11**, 95–112.

Ainley, D.G. and Schlatter, R.P. (1972) Chick raising ability of Adelie Penguins. *Auk* **89**, 559–566.

Amling, W. (1980) Exclusion of gulls from reservoirs in Orange County, California. *Vertebrate Pest Conf.* **9**, 4–6.

Anderson, D.W. and Gress, F. (1983) Status of a northern population of California Brown Pelicans. *Condor* **85**, 79–88.

Anderson, K.P. and Ursin, E. (1977) A multispecies extension to the Beverton and Holt theory of fishing, with accounts of phosphorus circulation and primary production. *Medd. Danm. Fisk. og Havunders.* **7**, 319–435.

Anderson, M., Gotmark, F. and Wicklund, C.G. (1981) Food information in the Black-headed Gull *Larus ridibundus*. *Behav. Ecol. Sociobiol.* **9**, 199–202.

Andrewartha, H.G. and Birch, L.C. (1954) *The Distribution and Abundance of Animals*. University of Chicago Press, Chicago.

Ankney, C.D. and MacInnes, C.D. (1978) Nutrient reserves and reproductive performance of female Lesser Snow Geese. *Auk* **95**, 459–471.

Anon (1983) Contaminants in marine top predators. *NERC Publications Series C* No. 23.

Appelquist, H., Asbirk, S. and Drabaek, I. (1984) Mercury monitoring: mercury stability in bird feathers. *Mar. Pollut. Bull.* **15**, 22–24.

Appelquist, H., Drabaek, I. and Asbirk, S. (1985) Variation in mercury content of guillemot feathers over 150 years. *Mar. Pollut. Bull.* **16**, 244–248.

Arnason, E. and Grant, P.R. (1978) The significance of kleptoparasitism during the breeding season in a colony of Arctic Skuas *Stercorarius parasiticus* in Iceland. *Ibis* **120**, 38–54.

Ashmole, N.P. (1963) The regulation of numbers of tropical oceanic birds. *Ibis* **103**, 458–473.

Ashmole, N.P. (1968) Body size, prey size and ecological segregation in five sympatric tropical terns. *Systematic Zoology* **17**, 292–304.

Ashmole, N.P. (1971) Seabird ecology and the marine environment. In *Avian Biology* Vol. 1, Farner, D.S., King, J.K. and Parkes, K.C. (eds.), Academic Press, New York, 224–286.

Ashmole, N.P. and Ashmole, M.J. (1967) Comparative feeding ecology of seabirds of a tropical oceanic island. *Peabody Mus. Natur. Hist., Yale Univ., Bull.* **24**, 1–131.

Bailey, E.P. (1982) Effects of fox farming on Alaskan islands and the proposed use of Red Foxes as biological control agents for introduced Arctic Foxes. *Bull. Pac. Seab. Group* **9**, 74–75.

Bailey, R.S. and Hislop, J.R.G. (1978) The effects of fisheries on seabirds in the northeast Atlantic. *Ibis* **120**, 104–105.

Baillie, S.R. (1987) *Seabird movements and mortality patterns in Western Europe*, British Trust for Ornithology, Tring.

Ball, N.J. and Amlaner, C.J. (1980) Changing heart rates of herring gulls when approached by humans. In *A Handbook on Biotelemetry and Radio Tracking*, Amlaner, C.J. and MacDonald, D.W. (eds.), Pergamon Press, Oxford, 589–594.

Baltz, D.M. and Morejohn, G.V. (1977) Food habits and niche overlap of seabirds wintering on Monterey Bay, California. *Auk* **94**, 526–543.

Baudinette, R.V. and Schmidt-Nielsen, K. (1974) Energy cost of gliding in herring gulls. *Nature* **248**, 83–84.

Becker, P.H. (1984) How a common tern (*Sterna hirundo*) colony defends itself against herring gulls. *Z. Tierpsychol.* **66**, 265–288.

Becker, P.H. and Erdelen, M. (1985) Coastal bird populations of the German Wadden Sea: Trends 1950–1979. In *Population and Monitoring Studies of Seabirds*, Tasker, M.L. (ed.), Proc. 2nd Int. Conf. Seabird Group, 11–12.

Bedard, J. (1969) Adaptive radiation in Alcidae. *Ibis* **111**, 189–198.

Bedard, J. (1976) Coexistence, coevolution and convergent evolution in seabird communities: a comment. *Ecology* **57**, 177–184.

Belopol'skii, L.O. (1961) *Ecology of Sea Colony Birds of the Barents Sea*. (Transl.) Israel Program for Scientific Translation, Jerusalem.

Bengtson, S.A. (1984) Breeding ecology and extinction of the Great Auk (*Pinguinus impennis*): Anecdotal evidence and conjectures. *Auk* **101**, 1–12.

Benton, C., Khan, F., Monaghan, P., Richards, W.N. and Shedden, C.B. (1983) The contamination of a major water supply by gulls (*Larus* spp.). A study of the problem and remedial action taken. *Water Research* **17**, 789–798.

Berger, M. and Hart, J.S. (1974) Physiology and energetics of flight. In *Avian Biology*, Vol. 4, Farner, D.S. and King, J.R. (eds.), Academic Press, New York, 416–477.

Berruti, A. (1983) The biomass, energy consumption and breeding of waterbirds relative to hydrological conditions at Lake St. Lucia. *Ostrich* **54**, 65–82.

Birkhead, T.R., Johnson, S.D. and Nettleship, P.N. (1985) Extra-pair matings and mate guarding in the common murre *Uria aalge*. *Anim. Behav.* **33**, 608–619.

Birkhead, T.R. and Furness, R.W. (1985) The regulation of seabird populations. In *Behavioural Ecology*, Sibly, R.M. and Smith, R.H. (eds.), Blackwell, Oxford, 145–167.

Bitman, J. and Cecil, H.C. (1970) Estrogenic activity of DDT analogs and polychlorinated biphenyls. *J. Agr. Food Chem.* **18**, 1108–1112.

Blake, B.F., Tasker, M.L., Hope-Jones, P., Dixon, T.J., Mitchell, R. and Langslow, D.R. (1984). *Seabird Distribution in the North Sea*. Nature Conservancy Council, Huntingdon.

Blus, L.T., Neely, B.S., Belisle, A.A. and Prouty, R.M. (1974) Organochlorine residues in Brown Pelican eggs: relation to reproductive success. *Environ. Pollut.* **7**, 81–91.

Bourne, W.R.P. (1976) Seabirds and pollution. In *Marine Pollution*, Vol. 6, Johnston, R. (ed.), Academic Press, London.

Bradley, P. and Monaghan, P. Audouin's Gull and the Chafarinas Islands Game Reserve. *Oryx*, in press.

Brockman, J.H. and Barnard, C.J. (1979) Kleptoparasitism in birds. *Anim. Behav.* **27**, 487–514.

Brough, T. and Bridgeman, C.J. (1980) An evaluation of long grass as a bird deterrent on British airfields. *J. Appl. Ecol.* **17**, 243–253.

Brown, C.R. (1984) Resting metabolic rate and energetic cost of incubation in Macaroni Penguins (*Eudyptes chrysolophus*) and Rockhopper Penguins (*E. chrysocome*). *Comp. Biochem. Physiol.* **77**, 345–350.

Brown, C.R. and Adams, N.J. (1984) Basal metabolic rate and energy expenditure during incubation in the Wandering Albatross (*Diomeda exulans*). *Condor* **86**, 182–186.

Brown, R.G.B. (1979) Seabirds of the Senegal upwelling and adjacent waters. *Ibis* **121**, 283–292.

Brown, R.G. (1980) Seabirds as marine animals. In *Behaviour of Marine Animals*, Vol. 4, Burger, J. (ed.), Plenum Press, New York, 1–39.

Bryant, D.M. (1979) Reproductive costs in the house martin (*Delichon urbica*). *J. Anim. Ecol.* **48**, 655–676.

Bryant, D.M. and Westerterp, K.R. (1983) Short-term variability in energy turnover by breeding house martins *Delichon urbica*: a study using doubly-labelled water ($D_2^{18}O$). *J. Anim. Ecol.* **52**, 525–543.

Bryan, G.W. (1984) Pollution due to heavy metals and their compounds. In *Marine Ecology*, Vol. 5, Kinne, O. (ed.), John Wiley, London.

Buckley, F.G. and Buckley, P.A. (1980) Habitat selection in marine birds. In *Behaviour of Marine Animals*, Vol. 4, Burger, J. (ed.), Plenum Press, New York, 69–112.

Bunyan, P.J. and Stanley, P.I. (1982) Toxic mechanisms in wildlife. *Reg. Toxicol. and Pharmacol.* **2**, 106–145.

Burger, J. (1980) The transition to independence and postfledging parental care in seabirds. In *Behaviour of Marine Animals*, Vol. 4, Burger, J. (ed.), Plenum Press, New York, 367–447.

Burger, J. (1981) Feeding competition between laughing gulls and herring gulls at a sanitary landfill. *Condor* **83**, 328–335.

Butler, P.J., West, N.H. and Jones, D.R. (1977) Respiratory and cardiovascular responses of the pigeon to sustained, level flight in a wind-tunnel. *J. exp. Biol.* **71**, 7–26.

Butler, P.J. (1980) The use of radio telemetry in the study of diving and flying in birds. In *A Handbook on Biotelemetry and Radio Tracking*, Amlaner, C.J. and MacDonald, D.W. (eds.), Pergamon Press, Oxford, 569–578.

Chabrzyk, G. and Coulson, J.C. (1976) Survival and recruitment in the Herring Gull *Larus argentatus*. *J. Anim. Ecol.* **45**, 187–203.

Clark, R.B. (1978) Oiled seabirds and conservation. *J. Fish. Reg. Bd. Can.* **35**, 765–678.

Connors, P.G. and Smith, K.G. (1982) Oceanic plastic particle pollution: suspected effect on fat deposition in Red Phalaropes. *Mar. Pollut. Bull.* **13**, 18–20.

Conover, M.R. (1983) Recent changes in Ring-billed Gull and California Gull populations in the Western United States. *Wilson Bull.* **95**, 362–383.

Conroy, J.W.H. (1975) Recent increases in penguin populations in the Antarctic and Sub-antarctic. In *The Biology of Penguins*, Stonehouse, B. (ed.), Macmillan, London.

Cooper, J. (1977) Food, breeding and coat colours of feral cats on Dassen Island. *Zool. Afr.* **12**, 250–252.

Coulson, J.C., Deans, I.R., Potts, G.R., Robinson, J. and Crabtree, A.N. (1972) Changes in organochlorine contamination of the marine environment of eastern Britain monitored by Shag eggs. *Nature* **236**, 454–456.

Coulson, J.C., Duncan, N. and Thomas, C. (1982) Changes in the breeding biology of the herring gull (*Larus argentatus*) induced by reduction in the size and density of the colony. *J. Anim. Ecol.* **51**, 739–756.

Coulson, J.C. (1983) The changing status of the Kittiwake *Rissa tridactyla* in the British Isles, 1969–1979. *Bird Study* **30**, 9–16.

Coulson, J.C. and Dixon, F. (1979) Colonial breeding in seabirds. In *Biology and Systematics of Colonial Organisms*. Larwood, G. and Rosen, B.R. (eds.), Academic Press, New York, 445–458.

Coulson, J.C. and Porter, J.M. (1985) Reproductive success of the kittiwake *Rissa tridactyla*: the roles of clutch size, chick growth rates and parental quality. *Ibis* **127**, 450–466.

Coulson, J.C. and Thomas, C. (1985a). Differences in the breeding performance of individual kittiwake gulls *Rissa tridactyla* (L.). In *Behavioural Ecology*, Sibly, R.M. and Smith, R.H. (eds.), Blackwell, London, 489–506.

Coulson, J.C. and Thomas, C.S. (1985*b*). Changes in the biology of the Kittiwake *Rissa tridactyla*: a 31-year study of a breeding colony. *J. Anim. Ecol.* **54**, 9–26.

Coulson, J.C. and Wooller, R.D. (1976) Differential survival rates among breeding Kittiwake gulls *Rissa tridactyla* (L.). *J. Anim. Ecol.* **45**, 205–213.

Coulter, M.C. and Risebrough, R.W. (1973) Shell thinning in eggs of the ashy petrel (*Oceanodroma homochroa*) from the Farallon Islands. *Condor* **75**, 254–255.

Cramp, S., Bourne, W.R.P. and Saunders, D. (1974) *The Seabirds of Britain and Ireland.* Collins, London.

Cramp, S. (1978) (ed.)*Handbook of the Birds of Europe, the Middle East and North Africa.* Vol. 1: *Ostrich to Ducks.* Oxford University Press, Oxford.

Cramp, S. (1985) (ed.) *Handbook of the Birds of Europe, the Middle East and North Africa.* Vol. 4: *Terns to Woodpeckers.* Oxford University Press, Oxford.

Cramp, S. and Simmons, K.E.L. (1983) (eds.) *Handbook of the Birds of Europe, the Middle East and North Africa.* Vol. 3: *Waders to Gulls.* Oxford University Press, Oxford.

Crawford, R.J.M. and Shelton, P.A. (1978) Pelagic fish and seabird interrelationships off the coasts of South West and South Africa. *Biol. Conserv.* **14**, 85–109.

Croxall, J.P. (1982) Energy cost of incubation and moult in petrels and penguins. *J. Anim. Ecol.* **51**, 177–194.

Croxall, J.P. and Prince, P.A. (1980) Food, feeding ecology and ecological segregation of seabirds at South Georgia. *Biol. J. Linn. Soc.* **14**, 103–131.

Croxall, J.P. and Prince, P.A. (1979) Antarctic seabird and seal monitoring studies. *Polar Record* **19**, 573–595.

Croxall, J.P. and Prince, P.A. (1982) A preliminary assessment of the impact of seabirds on marine resources at South Georgia. *Com. Nat. Fr. Recherch. Antarct.* **51**, 501–509.

Croxall, J.P. and Ricketts, C. (1983) Energy costs of incubation in the Wandering Albatross *Diomedea exulans. Ibis* **125**, 33–39.

Croxall, J.P., Ricketts, C. and Prince, P.A. (1984) Impact of seabirds on marine resources, especially krill, of South Georgia waters. In *Seabird Energetics*, Whittow, G.C. and Rahn, H. (eds.), Plenum Press, New York, 285–318.

Cushing, D.H. (1975) *Marine Ecology and Fisheries.* Cambridge University Press, Cambridge.

Daan, N. (1975) Consumption and production in North Sea cod *Gadus morhua*: an assessment of the ecological status of the stock. *Neth. J. Sea Res.* **9**, 24–55.

Daan, N. (1978) Changes in cod stocks and cod fisheries in the North Sea. *Rapp. et p-v. Reun. Cons. Int. Explor. Mer* **172**, 39–57.

Davis, R.W., Kooyman, G.L. and Croxall, J.P. (1983) Water flux and estimated metabolism of free-ranging Gentoo and Macaroni Penguins at South Georgia. *Polar Biol.* **2**, 41–46.

Day, R.H. (1980) The occurrence and characteristics of plastic pollution in Alaska's marine birds. Thesis, University of Alaska.

Diamond, A.W. (1978) Feeding strategies and population size in tropical seabirds. *Amer. Natur.* **112**, 215–223.

Drury, W.H. (1973) Population changes in New England seabirds. *Bird-Banding* **44**, 267–313.

Duffy, D.C. (1983) Competition for nesting space among Peruvian guano birds. *Auk* **100**, 680–688.

Dunnet, G.M., Ollason, J.C. and Anderson, A. (1979) A 28 year study of breeding Fulmars *Fulmarus glacialis* in Orkney. *Ibis* **121**, 293–300.

Dunn, E.H. (1979) Time-energy use and life history strategies of Northern seabirds. In *Conservation of Marine Birds of Northern North America.* Bartonek, J.C. and Nettleship, P.N. (eds.), *US Dept. Int. Wildl. Res. Rep.* **11**, 141–166.

Dyck, J. and Kraul, I. (1984) Environmental pollutants and shell thinning in eggs of the

Guillemot *Uria aalge* from the Baltic Sea and the Faroes, and a possible relation between shell thickness and sea water salinity. *Dansk Orn. Foren. Tidsskr.* **78**, 1–14.

Ellis, H.I. (1984) Energetics of free-ranging seabirds. In *Seabird Energetics*, Whittow, G.C. and Rahn, H. (eds.), Plenum Press, New York, 203–234.

Emlen, S.T. (1984) Co-operative breeding in birds and mammals. In *Behavioural Ecology*, 2nd edn., Krebs, J.R. and Davies, N.B. (eds.), Blackwell, London, 305–349.

Evans, P.G.H. and Waterston, G. (1976) The decline of the thick-billed murre in Greenland. *Polar Record* **18**, 283–293.

Evans, P.R. (1973) Avian resources of the North Sea. In *North Sea Science*, Goldberg, E.D. (ed.), Cambridge, Massachusetts.

Everson, I. (1977) *The living resources of the Southern Ocean*. FAO Southern Ocean Fisheries Survey Programme, Rome.

Evans, R.M. (1980) Development of behaviour in seabirds: an ecological perspective. In *Behaviour of Marine Animals*, Vol. 4, Burger, J. (ed.), Plenum Press, New York, 271–322.

Everson, I. and Ward, P. (1980) Aspects of Scotia Sea zooplankton. *Biol. J. Linn. Soc.* **14**, 93–101.

Feare, C.J. (1976a) The breeding of the Sooty Tern (*Sterna fuscata*) in the Seychelles and the effects of experimental removal of its eggs. *J. Zool., Lond.* **179**, 317–360.

Feare, C.J. (1976b) The exploitation of Sooty Tern eggs in the Seychelles. *Biol. Conserv.* **10**, 169–181.

Feare, C.J. (1984) Human exploitation. In *Status and Conservation of the World's Seabirds*, Croxall, J.P., Evans, P.G.H. and Schreiber, R.W. (eds.), ICBP, Cambridge.

Ferns, P.N., Macalpine-Levy, F.H. and Goss-Custard, J.D. (1980) Telemetry of heart rate as a possible method of estimating energy expenditure in the redshank *Tringa totanus* (L.). In *A Handbook on Biotelemetry and Radio Tracking*, Amlaner, C.J. and MacDonald, D.W. (eds.), Pergamon Press, Oxford, 595–602.

Fimreite, N., Bjerk, J.E., Kveseth, N. and Brun, E. (1977) DDE and PCBs in eggs of Norwegian seabirds. *Astarte* **10**, 15–20.

Fisher, J. (1952) *The Fulmar*. Collins, London.

Ford, R.G., Wiens, J.A., Heinemann, D. and Hunt, G.L. (1982) Modelling the sensitivity of colonially breeding marine birds to oil spills: guillemot and kittiwake populations on the Pribilof Islands, Bering Sea. *J. Appl. Ecol.* **19**, 1–31.

Frost, P.G.H., Siegfried, W.R. and Cooper, J. (1976) Conservation of the jackass penguin (*Spheniscus demersus* (L.). *Biol. Conserv.* **9**, 79–99.

Furness, B.L. (1983) Plastic particles in three procellariiform seabirds from the Benguela current, South Africa. *Mar. Pollut. Bull.* **14**, 307–308.

Furness, R.W. (1978a) Kleptoparasitism by Great Skuas (*Catharacta skua* Brünn) and Arctic Skuas (*Stercorarius parasiticus* L.) at a Shetland seabird colony. *Anim. Behav.* **26**, 1167–1171.

Furness, R.W. (1978b) Energy requirements of seabird communities: a bioenergetics model. *J. Anim. Ecol.* **47**, 39–53.

Furness, R.W. (1982) Competition between fisheries and seabird communities. *Adv. Mar. Biol.* **20**, 225–307.

Furness, R.W. (1983) *The Birds of Foula*. Brathay, Ambleside.

Furness, R.W. (1985a) Plastic particle pollution: accumulation by Procellariiform seabirds at Scottish colonies. *Mar. Pollut. Bull.* **16**, 103–106.

Furness, R.W. (1985b) Ingestion of plastic particles by seabirds at Gough Island, South Atlantic Ocean. *Environ. Pollut.* **38**, 261–272.

Furness, R.W. and Ainley, D.G. (1984) Threats to seabird populations presented by commercial fisheries. In *Status and Conservation of the World's Seabirds*, Croxall, J.P., Evans, P.G.H. and Schreiber, R.W. (eds.), ICBP, Cambridge, 701–708.

Furness, R.W. and Barrett, R.T. (1985) The food requirements and ecological relationships of a seabird community in North Norway. *Ornis Scand.* **16**, 305–313.

Furness, R.W. and Birkhead, T.R. (1984) Seabird colony distributions suggest competition for food supplies during the breeding season. *Nature* **311**, 655–656.

Furness, R.W. and Burger, A.E. (1987) Effects of energy constraints on seabirds breeding at high latitudes. *Proc. XIX Int. Orn. Congr. Ottawa.*

Furness, R.W. and Cooper, J. (1982) Interactions between breeding seabird and pelagic fish populations in the Southern Benguela region. *Mar. Ecol. Prog. Ser.* **8**, 243–250.

Furness, R.W. and Hislop, J.R.G. (1978) Diets and feeding ecology of great skuas *Catharacta skua* during the breeding season in Shetland. *J. Zool. Lond.* **195**, 1–23.

Furness, R.W. and Hutton, M. (1979) Pollutant levels in the Great Skua *Catharacta skua*. *Environ. Pollut.* **19**, 261–268.

Furness, R.W., Muirhead, S.J. and Woodburn, M. (1986) Using bird feathers to measure mercury in the environment: relationships between mercury content and moult. *Mar. Pollut. Bull.* **17**, 27–30.

Furness, R.W. and Todd, C.M. (1984) Diets and feeding of Fulmars *Fulmarus glacialis* during the breeding season: a comparison between St Kilda and Shetland colonies. *Ibis* **126**, 379–387.

Gambell, R. (1973) Some effects of exploitation on reproduction in whales. *J. Reprod. Fertil.* **19**, 533–553.

Gaston, A.J., Chapdelaine, G. and Noble, D.G. (1983) The growth of Thick-billed Murre chicks at colonies in Hudson Strait: inter- and intra-colony variation. *Can. J. Zool.* **61**, 2465–2475.

Gilman, A.P., Fox, G.A., Peakall, D.B., Teeple, S.M., Carroll, T.R. and Haymes, G.T. (1977) Reproductive parameters and egg contaminant levels of Great Lakes Herring Gulls. *J. Wildl. Manage.* **41**, 458–468.

Gilman, A.P., Hallett, D.J., Fox, G.A., Allan, L.J., Learning, W.J. and Peakall, D.B. (1978) Effects of injected organochlorines on naturally incubated Herring Gull eggs. *J. Wildl. Manage.* **42**, 484–493.

Girdwood, R.W.A., Fricker, C.R., Munro, D., Shedden, C.B. and Monaghan, P. 1985. The incidence and significance of Salmonella carriage by gulls (*Larus* spp.) in Scotland. *J. Hyg.* **95**, 229–234.

Goede, A.A. and de Bruin, M. (1984) The use of bird feather parts as a monitor for metal pollution. *Environ. Pollut. B.* **8**, 281–298.

Grant, G.S. and Whittow, G.C. (1983) Metabolic cost of incubation in the Laysan Albatross and Bonin Petrel. *Comp. Biochem. Physiol.* **74A**, 77–82.

Greig, S.A., Coulson, J.C. and Monaghan, P. (1983) Age-related differences in foraging success in the herring gull (*Larus argentatus*). *Anim. Behav.* **31**, 1237–1243.

Greig, S.A., Coulson, J.C. and Monaghan, P. (1985) Feeding strategies of male and female adult herring gulls (*Larus argentatus*). *Behaviour* **94**, 41–59.

Greig, S.A., Coulson, J.C. and Monaghan, P. Comparative foraging behaviour of 3 species of gulls at refuse tips. *J. Zool., Lond.* (in press).

Grenfell, B.T. and Lawton, J.H. (1979) Estimates of the krill consumed by whales and other groups in the Southern ocean: 1900 and the present. Unpublished manuscript.

Gress, F., Risebrough, R.W. and Sibley, F.C. (1971) Shell thinning in eggs of the common murre, *Uria aalge*, from the Farallon Islands, California. *Condor* **73**, 369–369.

Guillet, A. and Furness, R.W. (1985) Energy requirements of a Great white pelican (*Pelecanus onocrotalus*) population and its impact on fish stocks. *J. Zool., Lond.* **205**, 573–583.

Hamilton, W.D. (1971) Geometry for the selfish herd. *J. Theor. Biol.* **31**, 295–311.

Harrison, P. (1983) *Seabirds: An Identification Guide*. Croom Helm, Beckenham, UK.

Harris, M.P. (1984) *The Puffin*. Poyser, Calton.

Harris, M.P. and Osborn, D. (1981) Effect of a polychlorinated biphenyl on the survival and breeding of puffins. *J. Appl. Ecol.* **18**, 471–479.

Harvey, G.R., Steinhauer, W.G. and Miklas, H.P. (1974) Decline of PCB concentrations in North Atlantic surface water. *Nature* **252**, 387–388.

Hasegawa, H. (1984) Status and conservation of seabirds in Japan, with special attention to the Short-tailed Albatross. In *Status and Conservation of the World's Seabirds*, Croxall, J.P., Evans, P.G.H. and Schreiber, R.W. (eds.), ICBP, Cambridge, 487–500.

Hempel, G. (1978) North Sea fisheries and fish stocks—a review of recent changes. *Rapp. et p.v. Reun. Cons. Int. Explor. Mer* **173**, 145–167.

Hislop, J.R.G. and Harris, M.P. (1985) Recent changes in the food of young Puffins *Fratercula arctica* on the Isle of May in relation to fish stocks. *Ibis* **127**, 234–239.

Hockey, P.A.R., Cooper, J. and Duffy, D.C. (1983) The roles of coastal birds in the functioning of marine ecosystems in Southern Africa. *S. Afr. J. Sci.* **79**, 130–134.

Hoffman, W., Heinemann, D. and Wiens, J.A. (1981) The ecology of seabird feeding flocks in Alaska. *Auk* **98**, 439–456.

Hogg, E.H. and Morton, J.K. (1983) The effects of nesting gulls on the vegetation and soil of islands in the Great Lakes. *Can. J. Bot.* **61**, 3240–3254.

Horn, H.S. and Rubenstein, D.I. (1984) Behavioural adaptations and life history. In *Behavioural Ecology*, 2nd edn., Krebs, J.R. and Davies, N.B. (eds.), Blackwell, London, 279–300.

Hunt, G.L. Jr. (1980) Mate selection and mating systems in seabirds. In *Behaviour of Marine Animals*, Vol. 4, Burger, J. (ed.), Plenum Press, New York, 113–151.

Hunt, G.L. Jr. (in press) Marine bird biomass and food consumption in the south-eastern Bering Sea: a preliminary comparison with Antarctic waters. *4th Symp. Antarct. Biol.*, Wilderness RSA.

Hutchinson, L.V., Wenzel, B.M., Stager, K.E. and Tedford, B.L. (1982) Further evidence for olfactory foraging by Sooty Shearwaters and Northern Fulmars. In *Marine Birds: their Feeding Ecology and Commercial Fisheries Relationships*. Nettleship, D.N., Sanger, G.A. and Springer, P.F. (eds.), Proc. Pacific Seabird Group Symp. 1982, 72–77.

Hutton, M. (1981) Accumulation of heavy metals and selenium in three seabird species from the United Kingdom. *Environ. Pollut. A* **26**, 129–145.

Imber, M.J. (1975) Petrels and predators. *Int. Council Bird Pres. Bull.* **12**, 260–263.

Jeffries, D.J. and Parslow, J.L.F. (1976) Thyroid changes in PCB-dosed guillemots and their indication of one of the mechanisms of action of these materials. *Environ. Pollut.* **10**, 293–311.

Jermyn, A.S. and Hall, W.B. (1978) Sampling procedures for estimating haddock and whiting discards in the North Sea by Scottish fishing vessels in 1976 and 1977. *Cons. Int. Explor. Mer Pap. Res. D* **9**, 1–4.

Jones, E. (1977) Ecology of the feral cat, *Felis catus* (L.), (Carnivora: Felidae), on Macquarie Island. *Aust. Wildl. Res.* **4**, 249–262.

Jones, R. and Hislop, J.R.G. (1978) Changes in North Sea haddock and whiting. *Rapp. et. p.v. Reun. Cons. Int. Explor. Mer* **172**, 58–71.

Jordan, R. (1967) The predation of guano birds on the Peruvian anchovy (*Engraulis ringens* Jenyns). *Rept. Calif. Coop. Oceanic Fish. Invest.* **11**, 105–109.

Jordan, R. and Fuentes, H. (1966) Las poblaciones de aves guaneras y su situacion actual. *Inst. del Mar de Peru* **10**, 1–31.

Jouventin, P. (1975) Mortality parameters in Emperor Penguins *Aptenodytes*. In *The Biology of Penguins*, Stonehouse, B. (ed.), Macmillan, London, 435–446.

Kaftanovski, Yu. M. (1951) Birds of the murre group of the eastern Atlantic. Studies of the fauna and flora of the USSR. *Moscow Society of Naturalists* **28**, 1–170.

Kendeigh, S.C., Dol'nik, V.R. and Gavrilov, V.M. (1977) Avian energetics. In

Granivorous Birds in Ecosystems, Pinowski, J. and Kendeigh, S.C. (eds.), Cambridge University Press, Cambridge.

King, J.R. (1974) Seasonal allocation of time and energy resources in birds. In *Avian Energetics*, Paynter, R.A. (ed.), Publ. Nuttall orn. Club. No. 15, Cambridge, Massachusetts.

King, W.B. (1984) Incidental mortality of seabirds in gillnets in the North Pacific. In *Status and Conservation of the World's Seabirds*, Croxall, J.P., Evans, P.G.H. and Schreiber, R.W. (eds.), ICBP, Cambridge.

Kooyman, G.L., Davis, R.W., Croxall, J.P. and Costa, D.P. (1982) Diving depths and energy requirements of King Penguins. *Science* **217**, 726–727.

Krebs, C.J. (1978) *Ecology: The Experimental Analysis of Distribution and Abundance*, 2nd edn., Harper and Row, New York.

Kuroda, N. (1961) A note on the pectoral muscles of birds. *Auk* **78**, 261–263.

Lack, D. (1966) *Population Studies of Birds*. Clarendon Press, Oxford.

Lack, D. (1967) Interrelationships in breeding adaptation as shown by marine birds. *Proc. XIV Int. Orn. Congr. Oxford* 3–31.

Lack, D. (1968) *Ecological Adaptations for Breeding in Birds*. Methuen, London.

Lakhani, K.H. and Newton, I. (1983) Estimating age-specific bird survival rates from ring recoveries—can it be done? *J. Anim. Ecol.* **52**, 83–92.

Lasiewski, R.C. and Dawson, W.R. (1967) A re-examination of the relation between standard metabolic rate and body weight in birds. *Condor* **69**, 13–23.

Laws, R.M. (1977) The significance of vertebrates in the Antarctic Marine Ecosystems. In *Adaptations within Antarctic Ecosystems*, Llano, G.A. (ed.), Smithsonian Institute, Washington, DC.

Lifson, N. and McClintock, R. (1966) Theory of use of the turnover rates of body water for measuring energy and material balance. *J. Theoret. Biol.* **12**, 46–74.

Lockyer, C.H. (1976) Growth and energy budgets of large baleen whales from the Southern Hemisphere. *A. C. M. R. R. Sci. Consult. Mar. Mammals*, FAO, Rome.

Lundbeck, J. (1962) Biologisch-statistishe untersuchungen über die deutsche Hochseefischerei iv, 5: Die Dampferfischerei in der Nordsee. *Berichte der Deutschen Wissenschaftlichen Kommission fur Meeresforschung* **16**, 177–246.

MacArthur, R.H. and Wilson, F.O. (1967) *The Theory of Island Biogeography*. Princeton University Press, Princeton.

Mackintosh, N.A. (1973) Distribution of postlarval krill in the Antarctic. *Discovery Rep.* **36**, 95–156.

Manuwal, D.A. (1972) The population ecology of the Cassin's Auklet on southeast Farallon Island, California. Ph.D. thesis, University of California.

Marshall, A.J. and Roberts, J.D. (1959) The breeding biology of equatorial vertebrates: Reproduction of Cormorants (Phalacrocoracidae) at latitude 0°20'. *Proc. Zool. Soc. London* **132**, 617–625.

May, R.M., Beddington, J.R., Clark, C.W., Holt, S.J. and Laws, R.M. (1979) Management of multispecies fisheries. *Science* **205**, 267–277.

Menzel, D.W., Ryther, J.H., Hulbert, E.M., Lorenzen, C.J. and Corwin, N. (1971) Production and utilisation of organic matter in Peru coastal current. *Invest. Pesq.* **35**, 43–59.

Miles, J.A.R. and Shrivastav, J.B. (1951) Ornithosis in certain seabirds. *J. Anim. Ecol.* **20**, 195–200.

Mineau, P. (1982) Levels of major organochlorine contaminants in sequentially-laid Herring Gull eggs. *Chemosphere* **11**, 679–685.

Monaghan, P. (1979) Aspects of the breeding biology of herring gulls *Larus argentatus* in urban colonies. *Ibis* **121**, 475–481.

Monaghan, P. (1980) Dominance and dispersal between feeding sites in the herring gull (*Larus argentatus*). *Anim. Behav.* **28**, 521–527.

Monaghan, P. (1983) Gulls: populations and problems. In *Enjoying Ornithology*. T. and A.D. Poyser, Calton, England, 232–237.

Monaghan, P. (1984) Applied ethology. *Anim. Behav.* **32**, 908–915.

Monaghan, P. and Coulson, J.C. (1977) Status of large gulls nesting on buildings. *Bird Study* **24**, 89–104.

Monaghan, P., Shedden, C.B., Ensor, K., Fricker, C.R. and Girdwood, R.W.A. (1985) Salmonella carriage by herring gulls in the Clyde area of Scotland in relation to their breeding ecology. *J. Appl. Ecol.* **22**, 669–679.

Monaghan, P. and Zonfrillo, B. (1986) Population dynamics of seabirds in the Firth of Clyde. *Proc. Royal Soc. Edinburgh* (in press).

Montevecchi, W.A. and Piatt, J. (1984) Composition and energy contents of mature inshore spawning capelin (*Mallotus villosus*): implications for seabird predators. *Comp. Biochem. Physiol.* **78A**, 15–20.

Moors, P.J. and Atkinson, I.A.E. (1984) Predation on seabirds by introduced animals, and factors affecting its severity'. In *Status and Conservation of the World's Seabirds*, Croxall, J.P., Evans, P.G.H. and Schreiber, R.W. (eds.), ICBP, Cambridge, 667–690.

Morant, P.D., Cooper, J. and Randall, R.M. (1981) The rehabilitation of oiled Jackass Penguins *Spheniscus demersus*, 1970–1980. In *Proceedings of the Symposium on Birds of the Sea and Shore*, Cooper, J. (ed.), African Seabird Group, Cape Town.

Morris, R.J. (1980) Plastic debris in the surface of the South Atlantic. *Mar. Pollut. Bull.* **11**, 164–166.

Mougin, J.L. and Prevost, J. (1980) Evolution annuelle des effectifs et des biomasses des oiseaux Antarctiques. *Rev. Ecol.* **34**, 101–133.

Murphy, E.C., Day, R.H., Oakley, K.L. and Hoover, A.A. (1984) Dietary changes and poor reproductive performance in Glaucous-winged gulls. *Auk* **101**, 532–541.

Naarding, J.A. (1980) Study of the Short-tailed Shearwater *Puffinus tenuirostris* in Tasmania. *Nat. Parks Wildl. Ser. Tasmania*, 1–78.

Nagy, K.A., Siegfried, W.R. and Wilson, R.P. (1984) Energy utilization by free-ranging Jackass Penguins *Spheniscus demersus*. *Ecology* **65**, 1648–1655.

Nelson, J.B. (1978) *The Sulidae: Gannets and Boobies*. Oxford University Press, Oxford.

Nelson, J.B. (1967) Etho-ecological adaptations in the great frigate bird. *Nature* 214, 318.

Nelson, J.B. (1980) *Seabirds*. Hamlyn, London.

Nelson, J.B. (1983) Contrasts in breeding strategies between some tropical and temperate marine pelecaniformes. *Studies in Avian Biology* **8**, 95–114.

Nicholson, J.K. and Osborn, D. (1983) Kidney lesions in pelagic seabirds with high tissue levels of cadmium and mercury. *J. Zool., Lond.* **200**, 99–118.

Nisbet, I.T.C. (1973) Courtship feeding, egg size and breeding success in Common Terns. *Nature* **241**, 141–143.

Nuttal, P.A. (1984) Tick-borne viruses in seabird colonies. *Seabird* **7**, 31–41.

Ollason, J.C. and Dunnet, G.M. (1978) Age, experience and other factors affecting the breeding success of the fulmar *Fulmarus glacialis* in Orkney. *J. Anim. Ecol.* **47**, 961–976.

Ollason, J.C. and Dunnet, G.M. (1983) Modelling annual changes in numbers of breeding Fulmars, *Fulmarus glacialis*, at a colony in Orkney. *J., Anim. Ecol.* **52**, 185–198.

Olsthoorn, J.C.M. (1984) Aspects of the availability of breeding sites in a seabird colony. Unpubl. MSc. thesis, University of Aberdeen.

Parslow, J.L.F. and Jefferies, D.J. (1977) Gannets and toxic chemicals. *Brit. Birds* **70**, 366–372.

Parsons, J. (1975) Seasonal variation in the breeding success of the herring gull: an experimental approach to pre-fledging success. *J. Anim. Ecol.* **44**, 553–573.

Parsons, J. (1976) Nesting density and breeding success in the herring gull *Larus argentatus*. *Ibis* **118**, 537–547.

Partridge, L. (1978) Habitat selection. In *Behavioural Ecology*, Krebs, J.R. and Davies, N.B. (eds.), Blackwell, London, 351–376.

Pearson, T.H. (1968) The feeding biology of sea-bird species breeding on the Farne Islands, Northumberland. *J. Anim. Ecol.* **37**, 521–551.

Pennycuick, C.J. (1969) The mechanics of bird migration. *Ibis* **111**, 525–556.

Pennycuick, C.J. (1982) The flight of petrels and albatrosses (Procellariiformes), observed in South Georgia and its vicinity. *Phil. Trans. Roy. Soc. Lond. B* **300**, 75–106.

Perrins, C.M. and Birkhead, T.R. (1983) *Avian Ecology*. Blackie, Glasgow and London.

Pettit, T.N., Ellis, H.I. and Whittow, G.C. (1985) Basal metabolic rate in tropical seabirds. *Auk* **102**, 172–174.

Pettit, T.N., Grant, G.S. and Whittow, G.C. (1981) Ingestion of plastics by Laysan Albatross. *Auk* **98**, 839–841.

Piatt, J.F. and Nettleship, D.N. (1985) Diving depths of 4 Alcids. *Auk* **102**, 293–297.

Piatt, J.F. and Reddin, D.G. (1984) Recent trends and implications for Thick-billed Murres of the West Greenland salmon fishery. In *Marine Birds: their Feeding Ecology and Commercial Fisheries Relationships*, Nettleship, D.N., Sanger, G.A. and Springer, P.F. (eds.), Proc. Pacific Seabird Group Symp. 1982, 208–210.

Pierotti, R. (1983.)Gull-Puffin interactions on Great Island, Newfoundland. *Biol. Conserv.* **26**, 1–14.

Potts, G.R., Coulson, J.C. and Deans, I.R. (1980) Population dynamics and the breeding success of the Shag *Phalacrocorax aristotelis*, on the Farne Islands, Northumberland. *J. Anim. Ecol.* **49**, 465–484.

Ratcliffe, D.A. (1970) Changes attributable to pesticides in egg breakage frequency and eggshell thickness in some British birds. *J. Appl. Ecol.* **7**, 67–115.

Reilly, W.J., Forbes, G.I., Paterson, G.M. and Sharp, J.C.M. (1981) Human and animal salmonellosis in Scotland associated with environmental contamination. 1973–79. *Veterinary Record* **108**, 553–555.

Ricklefs, R.E. (1967) A graphical method of fitting equations to growth curves. *Ecology* **48**, 978–983.

Ricklefs, R.E. (1974) Energetics of reproduction in birds. In *Avian Energetics*, Paynter, R.A. (ed.), Publ. Nuttall Orn. Club No. 15, Cambridge, Massachusetts.

Ricklefs, R.E., Day, C.H., Huntington, C.E. and Williams, J.B. (1985) Variability in feeding rate and meal size of Leach's storm-petrel at Kent Island, New Brunswick. *J. Anim. Ecol.* **54**, 883–898.

Robinson, J., Richardson, A., Crabtree, A.N., Coulson, J.C. and Potts, G.R. (1967) Organochlorine residues in marine organisms. *Nature* **214**, 1307–1311.

Rochard, J.B.A. and Horton, N. (1980) Birds killed by aircraft in the United Kingdom, 1966–77. *Bird Study* **27**, 227–234.

Ryder, J.P. (1980) The influence of size on the breeding biology of colonial nesting seabirds. In *Behaviour of Marine Animals*, Vol. 4, Burger, J. (ed.), Plenum Press, New York, 153–168.

Salmonsen, F. (1979) Marine birds in the Danish Monarchy and their conservation. In *Conservation of Marine Birds of Northern North America*, Bartonek, J.C. and Nettleship, D.N. (eds.), US Dept. Int. Fish Wildl. Serv, 267–287.

Santander, H. (1980) *The Peru Current System 2: Biological Aspects*. UNESCO, Paris.

Schaefer, M.B. (1970) Men, birds and anchovies in the Peru current—dynamic interactions. *Trans. Amer. Fish. Soc.* **9**, 461–467.

Schneider, D. and Hunt, G.L. (1982) Carbon flux to seabirds in waters with different mixing regimes in the southeastern Bering Sea. *Mar. Biol.* **67**, 337–344.

Schneider, D. and Hunt, G.L. (1984) A comparison of seabird diets and foraging distribution around the Pribilof Islands, Alaska. In *Marine Birds: their Feeding Ecology and Commercial Fisheries Relationships*, Nettleship, D.N., Sanger, G.A. and Springer, P.F. (eds.), Proc. Pacific Seabird Group Symp. 1982. 86–95.

Schreiber, R.W. and Schreiber, E.A. (1984) Central Pacific seabirds and the El Nino Southern Oscillation: 1982 to 1983 Respectives. *Science* **225**, 713–716.

Shea, R.E. and Ricklefs, R.E. (1985) An experimental test of the idea that food supply limits growth rate in a tropical pelagic seabird. *Am. Nat.* **126**, 116–122.

Sherman, K., Jones, C., Sullivan, L., Smith, W., Berrieu, P. and Ejsymont, L. (1981) Congruent shifts in sandeel abundance in western and eastern North Atlantic ecosystems. *Nature* **291**, 486–489.

Smith, G.B. (1981) The biology of walleye pollock. In *The Eastern Bering Sea Shelf: Its Oceanography and Resources*, Vol. 2, Hood, D.W. and Calder, J.A. (eds.), Govt. Printing Office, Washington, DC.

Snow, P.W. (1965) The breeding of the Red-billed Tropic Bird in the Galapagos Islands. *Condor* **67**, 210–214.

Sobey, D. (1976) The effect of Herring Gulls on the vegetation of the Isle of May. *Trans. Bot. Soc. Edinburgh* **42**, 469–485.

Spaans, A.F. (1971) On the feeding ecology of the herring gull *Larus argentatus* Pont in the Northern part of the Netherlands. *Ardea* **59**, 73–188.

Spencer, R. (1966) Report in bird-ringing for 1965. *Brit. Birds* **59**, 441–491.

Steele, J.H. (1974) *The Structure of Marine Ecosystems*. Yale University Press, New Haven.

Stonehouse, B. (1960) The King Penguin *Aptendodytes patagonia* of South Georgia. 1. Breeding behaviour and development. *Falklands Dep. Surv. Sci. Rep.* 23.

Stowe, T.J. and Underwood, L.A. (1984) Oil spillages affecting seabirds in the United Kingdom, 1966–1983. *Mar. Pollut. Bull.* **15**, 147–152.

Summers, R.W. (1977) Distribution, abundance and energy relationships of waders (Aves: Charadrii) at Langesaan Lagoon. *Trans. Roy. Soc. S. Afr.* **42**, 483–494.

Swartz, L.G. (1966) Sea-cliff birds. In *Environment of the Cape Thompson Region, Alaska*, Wilimovsky, N.J. and Wolf, J.N. (eds.), United States Atomic Energy Commission, Oak Ridge.

Tanabe, S., Tanaka, H. and Tatsukawa, R. (1984) Polychlorobiphenyls, DDT, and Hexachlorocyclohexane isomers in the Western North Pacific ecosystem. *Arch. Environ. Contam. Toxicol.* **13**, 731–738.

Tasker, M.L., Hope-Jones, P., Blake, B.F. and Dixon, T.J. (1985) The marine distribution of the Gannet *Sula bassana* in the North Sea. *Bird Study* **32**, 82–90.

Taylor, K. and Reid, J.B. (1981) Earlier colony attendance by Guillemots and Razorbills. *Scott. Birds* **11**, 173–180.

Thomas, G.J. (1972). A review of gull damage and management methods at nature reserves. *Biol. Conserv.* **4**, 117–127.

Trivelpiece W. and Volkman, W.J. (1982) Feeding strategies of sympatric South Polar Skua *Catharacta maccormicki* and Brown Skuas *C.lonbergi*. *Ibis* **124**, 50–54.

Tucker, V.A. (1973) Bird metabolism during flight: evaluation of a theory. *J. Exp. Biol.* **58**, 689–709.

Tuck, L.M. and Squires, H.J. (1955) Food and feeding habits of Brunnich's Murre (*Uria lomvia lomvia*) on Akpatok Island. *J. Fish. Res. Bd. Can.* **12**, 781–792.

Turcek, F.J. (1969) On plumage quantity in birds. *Ekologia Polska* Ser. A **14**, 617–634.

Van Aarde, R.J. (1979) Distribution and density of the feral house cat *Felis catus* on Marion Island. *S. Afr. J. Wildl. Res.* **9**, 14–19.

Veen, J. (1977) Functional and causal aspects of nest distribution in colonies of the sandwich tern. *Behaviour Suppl.* **XX**, 1–193.

Vermeer, K. and Rankin, L. (1984) Influence of habitat destruction and disturbance on nesting seabirds. In *Status and Conservation of the World's Seabirds*, Croxall, J.P., Evans, P.G.H. and Schreiber, R.W. (eds.), ICBP, Cambridge, 723–736.

Walsberg, G.E. (1983) Avian ecological energetics. In *Avian Biology*, Vol. 7, Farner, D.S., King, J.R. and Parkes, K.C. (eds.), Academic Press, New York, 161–220.

Ward, P. and Zahavi, A. (1973) The importance of certain assemblages of birds as 'information-centres' for food finding. *Ibis* **115**, 517–534.

Warham, J. (1971) Body temperatures of petrels. *Condor* **73**, 214–219.

Warham, J. (1963) The Rockhopper Penguin *Endyptes chrysocane* at Macquarie Island. *Auk* **80**, 229–256.

Weathers, W.W., Buttemer, W.A., Hayworth, A.M. and Nagy, K.A. (1984) An evaluation of time-budget estimates of daily energy expenditure in birds. *Auk* **101**, 459–472.

Weathers, W.W. (1979) Climatic adaptation in avian standard metabolic rate. *Oecologia* **42**, 81–89.

Wiens, J.A. (1984) Modelling the energy requirements of seabird populations. In *Seabird Energetics*, Whittow, G.C. and Rahn, H. (eds.), Plenum Press, New York.

Wiens, J.A. and Innis, G.S. (1974) Estimation of energy flow in bird communities: a population bioenergetics model. *Ecology* **55**, 730–746.

Wiens, J.A. and Scott, J.M. (1975) Model estimation of energy flow in Oregon coastal seabird populations. *Condor* **77**, 439–452.

Williams, A.J. (1978) Mineral and energy contributions of petrels (Procellariiformes) killed by cats to the Marion Island terrestrial ecosystem. *S. Afr. J. Antarct. Res.* **8**, 49–53.

Williams, A.J., Siegfried, W.R., Burger, A.E. and Berruti, A. (1979) The Prince Edward Islands: a sanctuary for seabirds in the Southern Ocean. *Biol. Conserv.* **15**, 59–71.

Winkler, D.W. and Walters, J.R. (1983) The determination of clutch size in precocial birds. In *Current Ornithology*, Vol. 1, Johnston, R.F. (ed.), Plenum Press, New York, 33–68.

Wittenberger, J.F. and Hunt, G.L. Jr. (1985) The adaptive significance of coloniality in birds. In *Avian Biology*, Vol. VIII, Farner, D.S., King, J.R. and Parkes, K.C. (eds.), Academic Press, New York, 1–78.

Wolff, E.W. and Peel, D.A. (1985) The record of global pollution in polar snow and ice. *Nature* **313**, 535–540.

Wynne-Edwards, V.C. (1955) Low reproductive rates in birds, especially seabirds. *Proc. XI Int. Orn. Congr., Basel*, 540–547.

Wynne-Edwards, V.C. (1962) *Animal Dispersion in Relation to Social Behaviour*. Oliver and Boyd, Edinburgh.

Index

161